More than the Sum of the Parts

More than the Sum of the Parts

of the Parts

Complexity in Physics and Beyond

HELMUT SATZ

University of Bielefeld, Germany

OXFORD

UNIVERSITY PRESS

OXFORD
UNIVERSITY PRESS

Great Clarendon Street, Oxford, OX2 6DP,
United Kingdom

Oxford University Press is a department of the University of Oxford.
It furthers the University's objective of excellence in research, scholarship,
and education by publishing worldwide. Oxford is a registered trade mark of
Oxford University Press in the UK and in certain other countries

© Helmut Satz 2022

Published in the United States of America by Oxford University Press
198 Madison Avenue, New York, NY 10016, United States of America

British Library Cataloguing in Publication Data
Data available

Library of Congress Control Number: 2022930086

ISBN 978–0–19–286417–8

DOI: 10.1093/oso/9780192864178.001.0001

Printed and bound by
CPI Group (UK) Ltd, Croydon, CR0 4YY

The laws of physics are simple, but nature is complex.

Per Bak
How Nature Works, 1996

Contents

Preface

Much of our natural science is based on the supposition that the whole is the sum of its parts. This assumption has in fact worked amazingly well and has provided us with a body of scientific knowledge on which much of our modern world is based. In recent years, however, it has become more and more evident that there is an immense number of phenomena for which this assumption does not hold. For many years, we got along by saying "let us assume the state to be in equilibrium" or "let us assume motion without friction", and more. Today we see that all the neglected phenomena form a new field of study, complexity, which is as great or greater than that considered so far in our conventional natural science.

Moreover, there turns out to be a considerable degree of universality for complex phenomena: complexity is observed in a vast variety of phenomena in nature. In conventional physics, many concepts are applicable only to issues that arise there, and to not much more. On the other hand, the behavior observed for the onset of criticality, leading to correlations between even very distant constituents—this behavior arises not only in the study of magnetism in condensed matter physics, but as well in the cosmology of the early universe and in the formation of flocks of birds. Related patterns are found in economic developments, stock market fluctuations, population growth, the spread of diseases. And by now, critical behavior has also become a part of conventional physics.

Not surprisingly, this had led to the appearance of complexity theory as a new field of research, with new fundamental concepts, such as emergence, self-organization, bifurcation, and more. This book is not meant as an introduction to complexity theory—for that, we refer a number of excellent works listed in the bibliography. Our aim here is to illustrate a variety of different phenomena in nature for which conventional science cannot give a satisfactory account, and to show that they generally arise as a result of collective many-body interactions. In some cases, these could be codified into aspects of an emergent complexity theory, in others not. Another possible subtitle for the book would thus be "unconventional physics". My aim is more to describe, to point out, rather than to strive for a general theory. The many-faceted picture presented by the different phenomena will hopefully serve as a challenge to future scientists, and to them I leave the formulation of a possible science of complexity.

The presentation here will be restricted to phenomena encountered in nature, to questions addressed to natural science. We will not consider issues in sociology, politics or economy; these are clearly well outside the range of competence of a theoretical physicist, even if and when they lead to fluctuation patterns not unlike those in physics. Nevertheless, with complexity being such a novel subject, mathematicians, physicists, and others, can well provide essential contributions to its understanding and its future development. Crucial steps—critical behavior (Ken Wilson), the approach to chaos (Mitchell Feigenbaum) and self-organization (Per Bak), came from physicists. Different views of the subject come from different orientations, and mine in certainly that of a physicist. In any case, the field thus is far from closed, and so I dare to give this presentation, although the subject of the book overlaps only partially with my own field of work, the thermodynamics of strong interactions. In recent years, however, complexity has been found to play an increasing role also here, and besides this, I have, been in touch with the topic and its initiators for many years.

The development of physics is quite naturally associated with the names of those who brought about the crucial changes—Galileo, Newton, Maxwell, Boltzmann, Planck, Einstein and more. The new paradigms associated with the physics of complex systems have not yet made their pioneering inventors well-known to the general public. In line with showing how the study of complex systems is changing our thinking, it therefore seems natural to point out who started it all. My private, physics-oriented list starts with the three names already mentioned: Kenneth Wilson, Mitchell Feigenbaum and Per Bak; it continues with many more, of course. These three have pointed out the road in physics which we now follow, and although they are all no longer with us, I have known all of them personally. So in a way, this is my memorial for them.

The book is meant for a general audience, interested in the new perspectives opening up now in the study of systems consisting of many similar or identical constituents. It is not a treatise, but rather an attempt to convey the discovery that a great variety of such systems lead to novel behavior due to collective interactions of the parts, that the whole is more than the sum of the parts.

References for additional reading, general developments and further information are provided in the bibliography at the end of the book. In particular, I include here presentations of the status of the rapidly growing theory of complexity. Besides this list of general as well as specific references, I cite in addition at the end of several chapters some books or articles of particular relevance for the specific topic of that chapter. One topic which is quite closely related to the present work is the structure of animal swarms; this is indeed quite similar to that of many-body systems is physics. I have not included it here, since it is covered in detail in my recent book "The Rules of the Flock", which was published last year by Oxford University Press.

I had the pleasure of discussing different aspects of the topic with various colleagues, and so my thanks for stimulating remarks go to Andrzej Bialas (Krakow), Philippe Blanchard (Bielefeld), Paolo Castorina (Catania) and Frithjof Karsch (Bielefeld). And I much appreciate notes by Shaun Bullett (London) and Bob Doyle (Harvard). Without the interference of Corona I could perhaps have had the pleasure of discussing with them in person. I hope that the future will bring such possibilities back.

I dedicate this book to the memory of my wife Karin.

Bielefeld, November 2021
Helmut Satz

1

Introduction

Du siehst den Wald vor lauter Bäumen nicht.
(You don't see the forest because of all the trees.)

German proverb

Divide and Conquer

From cave drawings to space telescopes, mankind has always, in one way or another, tried to figure out the world in which we live. We want to understand what things are made of, how they function, what different forms they can take, and what the forces are that they experience. For the past two millennia and more, the physical sciences, with much help from mathematics, have developed an extremely successful approach to addressing and answering these questions, best summarized in the old Roman advice "divide and conquer." Instead of looking at the complex overall picture, we single out a small part of the whole and try to understand its workings. If we succeed, we then put many such parts together to arrive at an understanding of the larger scene. This philosophy started in ancient Greece, when the idea of atoms was introduced as the ultimate building blocks of matter, and the multi-faceted world was attributed to the different ways these constituents were put together. The approach of reduction to ultimate parts has, over the past centuries, led to a well-defined atomic structure, first with atoms made of nuclei and electrons, though the nuclei were then found to consist of nucleons (protons and neutrons), and these in turn of quarks. The interaction between the different constituents is mediated by electromagnetic as well as strong and weak nuclear forces, and in the past few decades, it has become possible to combine all the components and the corresponding forces into one unified description, the so-called *standard model* of elementary particle physics—the closest to a "world formula" that we have ever had. The only force which has so far resisted a final unification with those of the standard model is gravity. In spite of many attempts by some of the most prominent physicists, such a "theory of everything" is still missing. Very recently, in fact, it has been suggested that the reason for this might be that gravity is of a fundamentally different nature from the other forces; we shall come back to this truly new view of things later.

Philosophers in ancient Greece provided not only the start of reductionism—they also warned that there are limits to this way of thinking. This is perhaps best summarized by Aristotle, when he noted that the whole is more than just the sum of its parts. In the process of reduction to ultimate constituents some of the features of the whole will necessarily be lost, and it is not clear if the specific way in which we subsequently recombine things will lead back to what we started with. Reduction and recombination are the yin and yang of the world, complementary but opposing, and understanding one does not imply understanding the other.

The success of reductionism has for many years overshadowed the other side of the coin: if we have the building blocks, how can they be put together, and what can be built from them? The knowledge of the structure of the atom still left most of the behavior of bulk matter unexplained, just as the anatomy of a bird tells us little about the behavior of flocks of birds. In many forms of matter, constituents separated far from each other are completely uncorrelated, and idealized systems of this kind can indeed be taken apart and recombined, in order to account for much of the observed behavior. For them, and as we shall see, only for them, the whole is simply the sum of the parts.

The Onset of Complexity

At the transition points from one state of matter to another, for evaporation, melting, freezing and more, this reductionism ends; the system now refuses to be divided into independent subsystems—distant constituents are now connected, everything becomes correlated. Physicists, irritated by this complication, called it *critical behavior*. Today we realize that there are more and more phenomena that make sense only when systems of many constituents undergo such devious doings. A single atom cannot freeze or evaporate. Such phenomena signal the beginning of complexity; the whole is now definitely more than the sum of its parts.

It has thus become more and more evident that an understanding of the nature of elementary particles and the forces between them, the ultimate reductionist world formula, the "theory of everything," is not sufficient for a full understanding of the behavior of systems of very many such particles. The opposite approach, the combination of constituents to form complex systems, turns out to have its own distinct laws; moreover, these often depend very little on the nature of the constituents and their interactions, and they thus are generally quite universal. The magnetization of iron, the condensation of a gas, the formation of a galaxy, or even that of a flock of birds—these all lead to very similar structural patterns. In this sense, the truly new physics in the past 50 years is the universal science of collective behavior, dealing for the first time with systems that are no longer simply the sum of all the smaller subsystems. The formulation of a theory of critical behavior through renormalization, the scale invariant behavior of systems of all sizes, brought the Nobel Prize in Physics in 1982 to the American theorist, Kenneth Wilson. Some years before, in 1977, the Russian–Belgian theorist, Ilya Prigogine, had already received this Prize for showing that dissipative behavior could lead to new collective structures. The world is full of physical phenomena that only manifest themselves in many-body systems. Such phenomena are the topic of this book, and I want to address them in a way accessible to the general interested reader, with a minimum use of mathematics. As you will see, there are times when I have found it impossible to avoid mathematics completely; in the words of a former president of our university, it is the language that God uses when he wants to speak to humans. And as with most languages, even if you don't understand every word, you can often still follow the argument. The book is not meant as a scientific treatise, but rather as a narrative, telling the reader how the fascinating concepts of complexity emerged in our understanding of the physical world.

We should, of course, begin by defining what is meant by complexity. Unfortunately, this is harder than it seems, and there are various, not always compatible, definitions. Our starting point is obviously a system of many interacting simple constituents. If the behavior of that system is uniform, if any two subsystems even far apart show the same form of behavior, then we consider the overall system as simple, as opposed to complex. Examples of simple systems of this kind are dilute gases, liquids at rest, and crystals. However, they remain simple only if left alone, in equilibrium; introducing temperature gradients, fluid flow, friction, stress and more turns them into complex systems, and it is for precisely this reason that in traditional classical physics such effects were usually assumed to be negligible. Simple systems are generally found to follow deterministic laws in a way that allows unique predictions for the change of state under a slight change of control parameters. Increasing the volume of an ideal gas leads to a predictable change of its pressure. In simple systems, small causes lead to small effects. In complex systems, the individual constituents are generally also subject to deterministic laws, but collective effects result in unexpected new macroscopic behavior. A temperature change of one degree can turn water into ice.

We shall see in more detail that systems of many identical particles can be combined to form different structures, and the transition from one such structure to another leads to critical behavior. Moreover, the collective behavior of many particles can produce effective "emergent" forces, which cannot be reduced to a pairwise interaction between individual constituents. These two aspects in a way define the topic of this book: critical behavior and emergent structures in many-body systems.

In the past few decades, it has been found that complex systems also show regularities, and even obey general laws. The transfer of heat in liquids shows striking flow effects, from convection patterns to chaos. As already mentioned, the development of mathematical formulations for such phenomena has produced a remarkable universality, with diverse applications from turbulence in fluids to the evolution of animal populations. It is for this reason that the structure of complex systems is quite often independent of the nature and interaction of its constituents. And there is a limit to the regularities of complex systems: if many-body systems show completely irregular and unpredictable behavior, if even an infinitesimal change of parameters leads to completely unexpected large-scale effects, then we speak of chaos. Simple as that seems, chaos shares with complexity the lack of an unambiguous definition. How do you measure the absence of order?

Since complexity is a very new field of research, it deals with diverse phenomena which at first sight (and often also at second) appear rather uncorrelated. For this reason, it seems unavoidable to present in this book an assortment of topics in a rather unstructured way—the logical pattern putting everything "in order" is not yet in place, and remains a challenge to future science. For the time being, we are confronted with a variety of concepts—critical behavior, self-organized criticality, emergence, intermittency, scale-free behavior, chaos, turbulence and many more; these concepts are obviously related, but much of the time it is not clear how.

I should note here that there already exists a new field of research, called complexity theory. Its aim is the study of mathematical models that describe complex systems, in some cases successfully, in others not yet. We will touch on these efforts from time to time, but the essential aim here is to introduce a variety of patterns of complex behavior found in nature, whether or not we have a theoretical framework to account for them. We want to look at phenomena that are not amenable to a description by the traditional methods of standard natural science, and we want to see if we can somehow begin to understand their patterns, even if we don't yet have a theory. And to restrict the issue, we shall indeed deal with phenomena in nature—leaving side questions arising in politics, economics, the stock market and more, questions that involve complexity and that are addressed by complexity theory. We want to look at complexity in nature.

Hints of Novel Behavior

To introduce the field, we begin with an issue that has been with us for many years. In physical science, we have progressed from the atoms of antiquity to the periodic table of elements, to atomic structure, to the formation of nuclei out of protons and neutrons, and on to the quark substructure of these nucleons. In all of the two thousand years in which this reductionist understanding has evolved, we still have not arrive at a satisfactory answer to the simple question why *time* in our world, in history as well as in cosmology, has a well-defined direction and never runs backwards. Evidently this is in many ways a much more fundamental issue than the grand unification of quarks and leptons in elementary particle theory—but it is an issue which requires new ways of thinking as well as new empirical input. We shall see that time as we know it arises as a collective effect in many-body systems.

Complex systems, as we had noted, very often cannot be separated into uncorrelated subsystems: even distant constituents are somehow still connected. Next we therefore address the simplest possible case of the formation of connectivity: percolation. We shall show that with increasing density, even randomly distributed objects undergo critical behavior, from isolated entities to global connectivity. Besides the conventional critical behavior, magnetization, condensation and more, we have in the past few years discovered another, *geometric* form of criticality. In Asia, this has been present for millennia in the form of the game of Go, but its more general structure has appeared as percolation only some hundred years ago. In its poetic version, it deals with water lilies on a pond. How many randomly placed lilies do we need to have in order to allow an ant to cross the pond without getting its feet wet? Here as well, nature does make a jump, and as the density of lilies increases, suddenly the crossing becomes possible.

We shall then look in some detail at the nature of forces in the physical world. Our thinking here is formed by the gravitational force causing the apple to fall from the tree or holding the moon in orbit around the earth. In a similar way, the electric force binds electrons to the nucleus to form atoms, and the nuclear force binds protons and neutrons to obtain nuclei. A force seems to be something like an invisible spring stretched between two (or more) objects, pulling them towards each other, or—in the case of two like electric charges—repelling each other. In recent years, a different kind of force has come into consideration. If a hole is punched into a tire, the air rushes out, as if pushed by some invisible force. But there is no specific force on each molecule of outgoing air: only the entire system suffers a drive towards a thermodynamically favored state of being. In physics terms, the system undergoes a change, from the compressed and ordered state in the tire to the disordered state of molecules freely streaming into the surrounding air. Hence the fictitious agent is denoted as the *emergent force*, arising only in many-body systems as a collective phenomenon.

Next we turn to the critical behavior already alluded to several times. It implies that we know what normal behavior is. In physics, normal is that a small cause leads to a small effect, that a small change of the conditions leads to a small change of the measurable properties. If we lower the temperature a little, the density or the structure of the system will change a little. This is true almost everywhere; but around 100 degrees centigrade, water evaporates, turns into vapor, and around zero degrees, water freezes, becoming ice. This kind of behavior has been called punctuated equilibrium: for a large range of parameter values, nothing happens, and then suddenly there is a complete change. In the past century, it has become ever clearer that this effect called for a new kind of physics. The proven wisdom that "natura non facit saltus"—nature does not make jumps—here is simply wrong: nature does jump. For the mathematicians, it meant that smooth, analytical behavior suddenly became singular. For many years, for both physicists and mathematicians, critical meant undesirable.

In the classical forms of critical behavior, one always considered systems whose parameters, temperature, density and the like, were slowly varied by some outside operator. The Danish physicist, Per Bak, pointed out that in fact most systems in nature don't need this outside help—they evolve on their own towards a critical point. Outside of religion, no operator triggers an earthquake or a tsunami. As a result, self-organized criticality has become an important new field of research.

In mathematics, complexity has led to the reconsideration of what we mean by dimension. For millennia, one, two and three dimensions were our world, and relativity added time as a fourth. The study of structures retaining broken patterns at all scales—the famous coastline of Britain or Norway—led the French mathematician Benoit Mandelbrot to introduce fractality: there exist structures whose dimensions are not integral, and may lie between one and two, for example.

Another concept we had to give up was the idea that deterministic equations lead to unambiguous, unique results. The so-called logistic map, a recursion relation common to the evolution of populations, was shown to lead to bifurcation: for certain parameters, the result oscillated between two values, and these in turn continued to split further. Mitchell Feigenbaum showed that this pattern eventually results in unpredictability, in chaos.

As mentioned, the atomic structure of matter had already been proposed on philosophical grounds in ancient Greece. Two thousand years later, it was called into question: the idea of atoms could help as a calculational tool to understand certain regularities (such as the periodic table of elements), but did atoms really exist, with mass and size? In this sense, atoms were the quarks of the 19th century. Even some eminent physicists expressed doubts... The ultimate confirmation came from something known since antiquity and today called Brownian motion. In perfectly smooth media, water or air, visible macroscopic particles, pollen or dust, dance around in an erratic fashion, reflecting the random motion of the invisible constituents of these media, atoms or molecules.

The scale-independence found in many natural phenomena, leading for example to the observed Gutenberg–Richter distribution observed for earthquakes, continues surprisingly—and so far really without any clear explanation—in unexpected domains. The frequency of words used in human text of whatever language was also found to follow such a power-law in usage ranking, and if we partition the positive integers into their prime number components, the frequency of these components also obeys such a law. What is the origin of such universality?

The advent of quantum physics brought yet another form of collective behavior. At extremely low temperatures, in many-body systems, electrons as the quantum components lead to the formation of a fifth state of matter, in addition to solids, liquids, gases and plasmas. The new state, a kind of sleeping liquid (in physics terminology the Bose–Einstein condensate), gives rise to such striking phenomena as superconductivity (the vanishing of electrical resistance) and superfluidity (the vanishing of viscosity for fluids). Again, these are phenomena that only make sense for collective systems of many constituents—a single constituent cannot show such properties.

In conclusion, there are two distinct approaches to achieving an understanding of the world around us. In the first and oldest, we identify its ultimate building blocks, their structure, and the form of their interactions. In the second, we study the general patterns and regularities that govern the collective behavior of systems of very many such constituents. This second approach is the subject of this book, and we find that many of these "rules of collective behavior" are highly universal. In very similar forms, they hold for quarks and electrons, magnets and stars, insects and birds and much more. They describe the formation of galaxies, the appearance of turbulence and the evolution of animal populations. Complexity rules a truly amazing world.

2

The Flow of Time

Fugit irreparabile tempus
(Time flees irretrievably)

Virgil (70–19 BC)

Time Invariance in Physics

Time flows irretrievably from past to future, and we flow with it like a piece of wood in a stream. Humans have always known that one cannot reverse its direction. Its course is defined by stellar constellations, geological structures, fossils, written records, our own lives, our memories, and more. We know what *has* happened, but we do not know what *will* happen. We can define an order for past, present and future: there is an obvious *historical arrow of time,* from the unchangeable to the unpredictable.

It is therefore quite astonishing that until not so long ago, the foundations of physics were "invariant under time reversal," in the terminology of the field. The fundamental axioms, Newton's laws of mechanics, Maxwell's equations for electrodynamics, the equations of Schrödinger and Dirac for quantum systems, as well as Einstein's relativity theory: they all allow time to go backward as well as forward. Given the present state of things, all these laws *pre*dict the future and *post*dict the past on equal terms. The time of physics was the ticking of an ideal pendulum: sixty ticks define a minute, which has passed and which will pass. We are just at one point in an infinite sequence of equal points. So for centuries, physical science provided no help in assigning a direction to time. And even Einstein's combination of space and time in the theory of relativity was of no help: the fourth coordinate in space-time was still forward-backward symmetric and thus evidently not the time that measured the life of our grandparents. In physics, processes were always reversible, but in the real world, in our lives, most were irreversible.

And as far as the universe was concerned, it was thought to be eternal and unchanging, the stars above us were the *firmament*; they were always there in an unchanging space and thus also were of no help in assigning a direction to time… When Einstein noticed that his equations for the geometry of the universe allowed time to move forward as well as backward, allowing expansion as well as contraction for the world, he reluctantly added a term in order to destroy this freedom, to prevent temporal changes of the universe; it didn't work… and he subsequently considered this try as one of his biggest blunders.

The reason was that the cosmological view of time soon changed completely, when in 1929 the American astronomer Edwin Hubble showed that distant galaxies are in fact receding from us, that the universe is expanding. This meant, as had in fact been anticipated just before, on the basis of Einstein's equations by Georges Lemaitre in Belgium, that it must have started from a very dense initial state, in the *Big Bang*, and subsequently expanded to form our present world. Lemaitre, a Catholic priest as well as a physicist, noted that physics had provided a way for creation. In 1998, it was moreover shown that this expansion is even accelerating. So today we also have a clear *cosmological arrow of time*, and it points in the same direction as the historical one. Our universe once was young and then grew older, just as we do. Space and time are not equivalent: in space, displacements are possible in any direction, in time in only one direction: forward.

Cosmology, human history, as well as most religions, thus all insist on a well-defined direction of time. Is there a way to also make physics choose a direction? The philosopher of science, Hans Reichenbach, programmatically claimed that "there is no other way to solve the problem of time than the way through physics." Eventually, this indeed turned out to be feasible, but it required a new way of thinking, and even then, the arrow at first seemed to point in the wrong direction, so we have to go into a little more detail.

Joule's Experiment

The situation is perhaps understood most readily by looking at an experiment which the British physicist James Prescott Joule carried out around 1850. Joule was the son of a brewer and later on continued the brewery himself; so his interest in things like pressure and temperature were not totally academic. For his experiment, he took a thermally insolated container, divided into two compartments; one contained a gas, the other nothing—it was as close to a vacuum as possible. When he now removed the dividing wall between the two compartments, the gas rapidly streamed into the empty part and quickly distributed itself uniformly over the entire, now larger container (see Fig. 2.1).

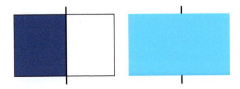

Fig. 2.1 Joule's experiment.

As long as one might wait: the gas as a whole never returned even briefly into its former compartment. The development, the expansion, was *irreversible*, it could not be undone. The arrow of time could not be turned around. On the other hand, if we were to perform such an experiment with just a single atom in the initial compartment, this atom would, after the removal of the dividing wall, be as likely to be in the old as in the new part of the container. And two atoms would with a probability of ¼ be back together in the old section, and as well as both in the new.

The breaking of the time symmetry, of time reversal invariance, must therefore somehow result from the fact that the gas consists of very many particles. With an increasing number of players, the simultaneous turn-about becomes ever more unlikely. We therefore expect that a well-defined direction of time appears only for systems consisting of many independent constituents, each moving its own way. It is perhaps the first indication in physics that the behavior of a many-body system is in principle conceptually different from that of its individual constituents.

The whole is indeed more than the sum of its parts, as Aristotle had noted more than two thousand years before. For a single atom, time does not matter, does not flow irretrievably; the flow of time is a collective phenomenon.

This means that God after all *does* play dice. To illustrate, let's consider a simpler situation, a case with nine boxes (see Fig. 2.2). Given nine balls, we then have *one* possible configuration, with one ball in each box. If we now double the size of the case to 18 boxes, with again nine balls, we get 48,620 different possible configurations. Only one of these is the initial starting case of the nine back in the initial section. If all configurations are equally likely, the chance for a full return to the starting state is 1:48,620.

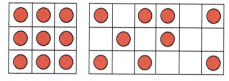

Fig. 2.2 Nine balls in nine boxes and one of the 48,620 possibilities of nine balls in 18 boxes.

A liter of a gas contains not nine, but some 10^{23} molecules; removing the divider thus results in a vast increase in the number of possibilities. The chances for a return are effectively zero: the likelihood that every particle retraces its proper path is vanishingly small. It appears that the directed flow of time is a collective property of many particles.

Entropy

The great Austrian physicist Ludwig Boltzmann created the basis for the physics of many constituents, *statistical mechanics*, and he turned the irreversible flow of the gas in Joule's experiment into an axiom of the field. He first explained that heat has a mechanical origin. It is a measure of the motion of the constituents of the system, so that heat indeed is a form of energy: that is the *first law of thermodynamics*. He then asked how many distinct configurations, *microstates*, the particles can form for a fixed total energy and volume. This gave him a new fundamental quantity of the field, the *entropy S*, defined as the logarithmic measure of the number W of possible microstates, compatible with the given *macroscopic* information, such as energy and volume. A given microstate is specified by the $3N$ positions and the $3N$ velocities of all its N constituents at a given moment, and for fixed energy and volume, there is evidently an immense number of such states. The position and velocity variables define what is denoted as the $6N$-dimensional *phase space*, and a given microstate corresponds to a point in this phase space.

Fig. 2.3 The tomb of Ludwig Boltzmann in Vienna, showing the definition of entropy.
(Photo courtesy of Oesterreichische Zentralbibliothek fuer Physik, Vienne, Austria)

As shown on the tomb of Boltzmann in Vienna (Fig. 2.3), the entropy is defined as $S = k \log W$. The proportionality constant k, the Boltzmann constant, together with the speed of light c, the gravitational constant G

and the Planck constant h, is one of the four basic constants of nature. It is the only one of the four which deals with many-body systems and has nothing to do with the interaction of individual particles. It is perhaps justified to say that (pre-relativity) classical physics rests on the shoulders of three giants: Isaac Newton for mechanics, James Clerk Maxwell for electrodynamics and Ludwig Boltzmann for statistical mechanics.

In the above example with the balls, $W = 1$ and hence $S = 0$ for the nine in the initial case, while the doubled case gives $S = 4.7\ k$. For a constant number of balls, the entropy thus increases with volume size. And so Boltzmann proposed what is today called the *second law of thermodynamics*: in the evolution of an isolated system the entropy never decreases; it either increases, or it remains constant if it is already at its maximum. If we give the medium a chance to increase its entropy by expanding into a larger volume, it always does so. Hence in the Joule experiment we have just that: the expansion of a gas at fixed total energy, at constant temperature, results in an entropy increasing with the overall volume. Although we have here illustrated the second law of thermodynamics for the case of a rather mundane example, it is perhaps the most profound statement in all of physics. The famous English physicist and astronomer Sir Arthur Eddington warned his colleagues: *If your favorite theory is contradicted by observation—well, these experimentalists do bungle things sometimes. But if your theory is found to be against the second law of thermodynamics, I can give you no hope; there is nothing for it but to collapse in deepest humiliation.*

We saw above that given a larger volume, the very orderly state of nine balls in nine adjacent boxes leads to a very low entropy and hence becomes very unlikely. This is in fact a very general feature: the more ordered a system is, the lower in general is its entropy. So the evolution in time of a system left to itself tends to decrease its order, increase its disorder or randomness. We could call this the *entropic direction of time*.

According to the cosmologists after Lemaitre and Hubble, our universe has been expanding ever since the Big Bang; as mentioned, this defines the cosmological arrow of time. According to statistical mechanics, this must also imply an increasing entropy. The very early universe must then have had a very much lower total entropy than our present one; quite a number of clever scientists have arrived at this conclusion.

There is a catch, however, as most of them noted. The early universe was, as far as we know today, a very hot gas of elementary particles, and such a system is, according to conventional statistical mechanics, already in a state of maximum entropy. Our present world, with galaxies, stars, planets and cosmologists, is certainly not at maximum entropy. If we were to turn everything around us into a gas of free protons and neutrons, the entropy would considerably increase. The entropy was maximal at the beginning, but it is not maximal now. Something seems not to match, the evolution of the universe appears to violate the entropic arrow of time, as given by statistical mechanics.

Why Is the Entropy of the Universe not Maximal?

Today, we believe to have the solution to this puzzle. The basic idea is due to the American physicist and cosmologist, David Layzer, of Harvard University. At the start of the Joule experiment, the gas is in equilibrium in the initial volume; its entropy there is maximal for the given volume. After the removal of the divider, the gas rushes out, expands and equilibrates, so that some time later it is again in equilibrium, in the larger volume and with greater entropy. In the intermediate stage,

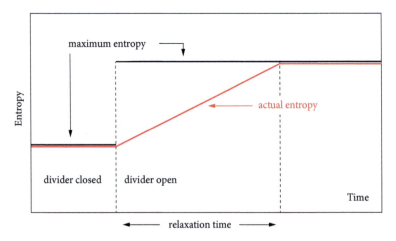

Fig. 2.4 The evolution of entropy in the Joule experiment.

the entropy rises from the initial (small volume) value toward the larger (large volume) value (see Fig. 2.4). In this intermediate stage, during the relaxation time, the system is not in equilibrium and the value of its actual entropy lies still below the value possible for the larger volume. The actual entropy is not yet maximal, because the gas particles are preferentially flying in the direction of the new volume, rather than in a direction orthogonal to this. In this *relaxation* stage, we thus have a certain order, which is eventually destroyed by collisions between the molecules, restoring isotropy and equilibrium.

In such a picture, and in fact more generally, order is the difference between the actual and the maximally possible entropy. It disappears when the entropy does become maximal—we then have complete disorder. So order becomes a lack of disorder, not the other way around, as we might think. But for the given system, disorder, maximum entropy, is unique, while there are many different kinds of order, from snowflakes to crystals.

The relaxation time is determined on the one hand by the strength of the interaction between the molecules: equilibration is the process in which the constituents eliminate differences through collisions. On the other hand, it depends on the density of the medium: if it is very dilute, there are only few collisions. Since expansion causes the density to decrease, the relaxation time increases. An expanding system is thus subject to a competition between the relaxation time and the rate of expansion. If the system expands rapidly enough, relative to the relaxation time, the actual entropy—although it is continuously increasing—falls more and more below the maximum possible value. We thus have a situation in which the entropy of the system continuously increases, in accord with the second law of thermodynamics, but at the same time, the difference between actual and maximum entropy is also growing: order is also continuously increasing, see Fig. 2.5. This scenario is the explanation that David Layzer (Fig. 2.6) suggested around 1975 for the seemingly opposite directions of the time evolution in cosmology and in statistical mechanics.

Statistical mechanics requires an increasing entropy, cosmology an increasing order. We see here that both can happen at the same time, starting from a state of maximum entropy at the time of the Big Bang.

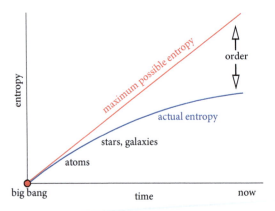

Fig. 2.5 Layzer's scenario for the evolution of the Universe.

Fig. 2.6 David Layzer (1925–2019).
(Photo courtesy of Jean Layzer)

The subsequent expansion of space was too rapid to allow the initial hot plasma of elementary particles to remain in equilibrium. Thus the temporal evolution into a larger volume led to an actual entropy which on one hand increased, and on the other fell more and more behind the maximum possible, which grew even faster. In this way, both order and disorder increased simultaneously, contrary to our common experience...

We should note that this picture does not tell us anything about the kind of order that can arise in the course of time in this evolution. That depends on the nature of the interactions between the constituents of the medium in question; we shall return to this issue in a subsequent chapter.

Before going on, we should note that irreversibility, the existence of an intrinsic arrow of time in specific physical phenomena, has been discussed extensively also on a more microscopic level. We have here addressed the issue in terms of the fundamental laws of nature. Such a discussion does not include non-equilibrium phenomena such as friction: the sliding motion of an object stopped by friction can evidently not be reversed. Similarly, self-organizing events such as the death of living beings or the decay of a radiative state are also irreversible. In quantum theory, the interaction of radiation and matter also tends to act as path erasure. For a discussion of these and similar phenomena, we refer to the quoted literature.

Further reading

The mentioned scenario for the direction of time is discussed in more detail in:

- D. Layzer, *Cosmogenesis: The Growth of Order in the Universe*, Oxford University Press, 1990
- D. Layzer, *The Arrow of Time*, Scientific American, 1975
- R. O. Doyle, *The Origin of Irreversibility*, The Information Philosopher 2014

The role of time in physics, including microscopic mechanisms for irreversibility, is discussed more generally in:

- P. Frampton, *Did Time Begin? Will Time End?*, World Scientific, Singapore 2009
- H. Reichenbach, *The Direction of Time*, U. California, Berkeley 1971 and Dover 1999.
- D. Zeh, *The Physical Basis for the Direction of Time*, Springer, Berlin 1989.

3

Global Connections

And God said, Let the waters under the heaven be gathered together unto one place, and let the dry land appear, and it was so.
The Bible, *Genesis 1.9*

Throwing Coins

The simplest possible model of a many-body system is probably obtained by randomly throwing identical coins onto a table of area A, allowing total or partial overlap. In the course of time this will produce on the table not only isolated single coins, but also clusters of several, partially overlapping coins, and the size of these clusters will grow with the number of deposited coins. At first, the rise is linear: the average cluster size G is proportional to the number N of thrown coins, $G = x N$, with an increase rate specified by the proportionality constant x. To eliminate the specific table size, we divide both sides by A and thus obtain

$$g = x n, \tag{3.1}$$

with n denoting the average density of coins and g the average cluster size relative to that of the table. Fig. 3.1a shows such distribution. Increasing the coin density makes it more and more likely that successive coins increase the size of existing clusters, leading to a growing increase of the average cluster size g, see Fig. 3.1b.

As we continue the game, at a certain specific density n_c the addition of a single further coin will cause the largest cluster to establish a global connection, a link between opposite sides of the table. Up to this point,

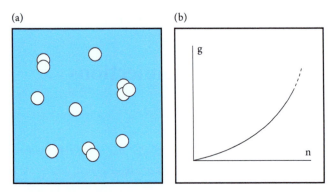

Fig. 3.1 (a) early coin distribution; (b) average relative cluster size g vs. coin density n.

we had islands of coins in a table sea; from now on, we have a multitude of lakes in a land of coins; the lakes can still contain islands. A small stimulus, one coin, thus leads to a fundamental change of structure, turning a world of islands in the sea into a world of lakes in a land region; see Fig. 3.2. For densities below n_c, a fish can swim from one side to the other, but a mouse cannot traverse the table without getting wet. Above the density n_c the situation is reversed. The value $n = n_c$ thus constitutes a critical point of the density. Above and below this value, small changes have small effects, but at $n = n_c$ one world turns into another.

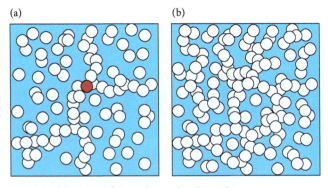

Fig. 3.2 (a) onset of percolation by the red coin; (b) configuration above the percolation point.

Percolation Theory

In mathematics, the onset of uninterrupted connection between the opposite sides of the table is called percolation, since from this point on a passage become possible (*percolare = pass through*). Percolation theory forms a field of mathematics, and it allows a (numerical) determination of the critical density. One thus obtains

$$n_c = \frac{1.13}{\pi r^2},$$

(3.2)

where r is the radius of the coin, so that πr^2 is its area. At the critical point, the total additive area of all thrown coins thus exceeds that of the table by some 10%; that is the consequence of the overlap of numerous coins. As a result, at $n = n_c$ about 32% of the table remains empty.

If we now continue throwing coins, eventually the entire table will be covered; the behavior of g as a function of the density n thus has the form shown in Fig. 3.3a. Of particular interest here is the rate of increase of the cluster size. Initially, for a small number of coins, g grows linearly, but then it begins to grow faster and faster. At the percolation point, a single coin suffices to create a cluster of the dimension of the whole table. There, at $n = n_c$, the rate of growth $x = (\Delta g/\Delta n)$ reaches its maximum value, which becomes infinite in the limit of an infinitely large system ($N \to \infty$, $A \to \infty$, $n = N/A$ fixed), see Fig. 3.3b.

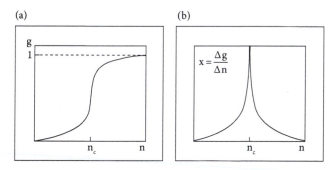

Fig. 3.3 (a) relative cluster size g vs. coin density n; (b) growth rate x of the cluster size vs. coin density n.

For mathematicians, critical behavior means that some observables diverge, i.e. become infinite. Percolation thus becomes a *critical process*; close to the critical density, the rate of growth of the cluster size leads to

$$X = \frac{\Delta g}{\Delta n} \sim \frac{1}{|n_c - n|^{\alpha}} \to \infty, \tag{3.3}$$

Where $|x|$ denotes the absolute value of x; x diverges when we approach n_c from above as well as from below. We thus find a divergence of the form 1/0, which can be further specified with the help of the so-called *critical exponent* α. In the linear region, for small densities n, the cause, an increase in density, and the effect, the growth of the relative cluster size g, are comparable. In the critical region, a tiny microscopic cause, one more coin, produces great macroscopic effects, the appearance of the global land bridge of coins. In this sense, percolation is indeed a form of complexity.

At this point, it seems natural to ask how critical behavior can arise for a system of apparently non-interacting objects, the randomly thrown coins. Some reflection will show that indeed the size of the coins plays the role of the interaction: the size of a cluster is crucially determined by the size of the individual coins. This most clearly seen in equation (3.2), in which the critical density is seen to be a universal number in terms, the coin size.

Site Percolation

As mentioned, percolation schemes apply to numerous physical situations. One quite striking example is given by the spreading of forest fires, described in detail in the cited book by Stauffer and Aharony. We briefly sketch the idea here; it is based on site percolation, in contrast to the continuum case we have considered so far, and more like the game of Go. A connected cluster is now defined as consisting of adjacent occupied sites (nearest neighbors only, not diagonal), and percolation occurs if a cluster connects the opposite sides of the lattice. For simplicity, we consider a plane square lattice, each site of which may or may not be occupied.

We denote the probability that a square is occupied by p, so that $(1 - p)$ applies to empty sites. Actual forests have $p < 1$. Two typical configurations are illustrated in Fig. 3.4; in Fig. 3.4b, the trees without any nearest neighbors are marked by black circles.

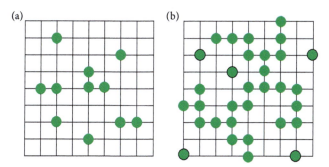

Fig. 3.4 Dilute (a) and percolating (b) site configurations

We start from Fig. 3.4a, with the green circles denoting trees; the rest is empty. Now assume that lightning strikes a tree and sets it on fire; we mark it red, see Fig. 3.5. To study the result, we consider successive time steps ("sweeps") for the lattice, starting with tree x being ignited in time step 1. If that tree has one or more nearest neighbors, such as the one marked x, these will also catch fire, while tree x now is burnt out and marked black (time step 2). In time step 3, the process continues; since the two neighbors of x do not have any nearest neighbors, they burn out and that is the end: after time step 3, only three trees are destroyed, the majority remains intact.

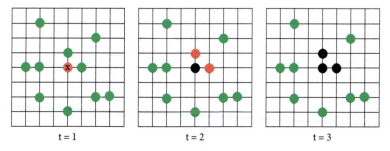

Fig. 3.5 The time step evolution of the fire in a dilute forest.

On the other hand, given configuration 3.4b, after sufficiently many time steps, all but four trees are destroyed (those marked by black circles in Fig. 3.4b), basically the forest is finished.

It is thus evident that the degree of survival of the forest depends on whether or not there is site percolation. In the limit of a large two-dimensional lattice, the critical percolation density is (numerically) found to be $p_c = 0.59 + / - 0.01$. As a result, we can predict that for $p < p_c$, the forest fire will end after a finite time, with the bulk of the forest remaining unharmed. On the other hand, for $p > p_c$, the fire will continue until essentially the entire forest (except for some isolated trees) is destroyed.

For completeness we note that besides the two forms considered here, continuous and site percolation, a third form is defined through bond percolation. The coins (or equivalent) which were randomly placed in the continuous case and on the lattice sites for site percolation are now positioned on the links or bonds of the lattice. This is the only one of the three which can be solved analytically, and the percolation density is found to be $p_c = 0.5$.

Further Reading

Percolation was first introduced in

S. R. Broadbent and J. M. Hammersley, *Percolation Processes*, Mathemat. Proc. Cambridge Philosophical Society 53 (1957) 629.

The standard introduction and text is perhaps

D. Stauffer and A. Aharony, *Percolation Theory*, Taylor & Francis, London 1985.

Later and more formal presentations are:

B. Bollobas and O. Riordan, *Percolation*, Cambridge University Press 2021.

G. Grimmet, *Percolation*, Springer, Heidelberg 1999

H. Kesten, *What is Percolation?*, Notices American Math. Society 53 (2006) 57.

4

The Nature of Forces

Ich bin ein Teil von jener Kraft, die stets das Böse will, und stets das Gute schafft.

(I am a part of that force which always wants evil but always creates good.)

Johann Wolfgang von Goethe, *Faust I*

Falling Stones

Among the forces we encounter in nature, gravity is certainly the most prominent. It defines the weight of our bodies, and it limits how high we can jump. We do work against it when lifting a stone, and gravity makes the stone fall back down when we let it go. The realization that the same force that causes an apple to fall from a tree also holds the earth in orbit around the sun formed the beginning of modern physics. Newton's law, stating that two bodies of masses m and M are attracted by the force of gravity inversely proportional to their square of their separation distance r,

$$F_g = G \, \frac{mM}{r^2} \tag{4.1}$$

provided us with the prototype picture of forces in nature; the proportionality constant G, the universal gravitational constant, is the same wherever masses attract each other, in outer space as well as on earth or in the solar system. The equation (4.1) allows us to calculate the trajectory of cannon balls, the path of the moon around the Earth or that of the Earth around the sun, and we use it today to determine the power of rocket engines to reach outer space. The masses m_i in the equation (4.1) represent their resistance against being moved, and there is no smallest or largest value for such a mass. If in some distant galaxy there is a star encircled by planets (and there are presumably billions of them), the motion of these planets is governed by equation (4.1). It is truly universal.

Let us recall at this point that the effect of a force is described by Newton's law, stating that $F_m = ma$, where a is the acceleration experienced by the mass m. Combining this with the force of gravity (4.1) we obtain

$$a = GM/r^2 \qquad (4.2)$$

Galileo Galilei's celebrated result: the motion of a falling object on Earth depends only on the mass M of the Earth and its radius r, giving $g = 9.8$ m/s^2 for the acceleration of gravity. The mass of the object itself does not enter, everything falls the same way (ignoring air resistance, of course).

The next interaction to appear in physics was the electric force. If one suspended two lightweight balls separated by a distance r and then touched each of them with an amber rod rubbed with a wool cloth, they repelled each other. Using an amber rod on one and a glass rod on the other led to attraction. The strength of attraction or repulsion depended on the amount rubbing on the rod, and it was found to vary as the inverse square of the separation. The result of this observation was Coulomb's law, as electric counterpart of Newton's law:

$$F_q = k_e \frac{q_2 q_1}{r^2}, \qquad (4.3)$$

where q_i measured the amount of electricity transferred by the touch of the rod; it could be positive or negative (amber or glass), and the resulting force is repulsive for like signs and attractive for unlike. It subsequently turned out, however, that there was one big difference between the two laws: the amount of electricity was given in terms of basic electric charges. Unlike mass, electricity had a smallest unit, $+e$ or $-e$, and the values q_i above were multiples of these basic charges. So the electric interaction was given as a fundamental force between elementary electric charges,

$$F_e = k_e \frac{e_2 e_1}{r^2} \tag{4.4}$$

The Coulomb constant k_e here plays the role of the gravitation constant G in equation (4.1) and is also universal. Equation (4.4) led physics back to its start in ancient Greece: all matter is made of some smallest possible units ("atoms"), coupled by universal forces. So it seems only natural that there is a smallest electric charge. Gravity, on the other hand, is also universal, but there is no smallest unit. And in contrast to falling objects, the motion of a massive electric charge does depend on its mass.

Subsequent studies led to two further forms of force. The nuclear or strong interaction held nucleons (protons and neutrons) together to form nuclei. Its range is much shorter than the inverse distance law of the above two, but again there are smallest possible "charges"—the basic interaction partners are nucleons. And when two protons get close enough to each other, their nuclear attraction is by many orders of magnitude stronger than the electric repulsion. In the present theory of strong interactions, quantum chromodynamics, the nucleons in turn become composites of quarks, making these the basic entities. The functional form of the force changes, but the structure of elementary charges interacting by fundamental forces remains. It also does so for the fourth and last form of interaction, the weak interaction governing radioactive decays. It attributes a basic lepton charge to electrons, muons and neutrinos.

So if we want to attribute the structure of matter to elementary constituents interacting through fundamental forces, we find that three of the four interactions so far observed in nature, strong, electromagnetic and weak, follow this pattern. Gravity, in contrast, does not provide us with a basic charge. It is also considerably weaker than the others: the electric repulsion between two protons is by a factor 10^{42} stronger than their attraction by gravity. And at still closer distances, the nuclear attraction is yet many orders of magnitude larger.

This means that on the one hand, our present understanding of gravity does not provide us with a fundamental gravitational "charge," and on the other hand, we have no way to test gravity alone on the level of single elementary particles. We expect that two protons, or a proton and an antiproton, are attracted to each other by gravity according to Newton's law, equation (4.1). But we have no way to test this, since at short distances their nuclear and at larger distances their electromagnetic interaction are overwhelmingly larger.

As already mentioned, we have today the "standard model," unifying strong and weak nuclear forces with those of electromagnetism, albeit with many constants. But gravity simply does not seem to fit into such a scheme. Attempts of resolve this remaining enigma are continuing through the world, pursued by numerous brilliant scientists, though so far without success.

Collective Gravity

There is another aspect in which gravity differs basically from all other forces. A system of many positive (or negative) electric charges is unstable—repulsion will drive it apart. Nuclear forces are very short-ranged, so that at very close distances, they can overcome this repulsion and allow the formation of nuclei, but only up to certain size. Eventually the repulsion wins, and so the size of nuclei is intrinsically limited. In a system of positive and negative charges in equal numbers, shielding creates finite size regions of electric neutrality; seen from afar, such a system is electrically neutral, so that the range of the electromagnetic force is limited for stable systems.

In contrast, gravity is additively attractive: a large system of many masses looks from the outside like one big mass; there is no repulsion. This is particularly relevant for galaxies—cosmic systems consisting of millions or even billions of stars. The motion of stars in such a galaxy is governed by Kepler's rule: they orbit as if they were moving around one big mass, made up of all the stars further in. In other words, gravity, and only gravity, does not allow a distinction between a single massive source or a collection of many individual masses. It is this feature which has led to the discovery of a striking new form of matter, observed through a detailed study of the motion of stars in galaxies. It is a form of matter which until today has remained completely enigmatic—we have no idea what its origins are.

THE NATURE OF FORCES 31

Dark Matter

Initially, Kepler's rule specified the orbital velocity v of planets around the sun: it arises by equating the centrifugal force of the orbiting planet to the gravitational attraction of the sun, giving

$$v^2 = \frac{GM}{d}, \tag{4.5}$$

where M denotes the mass of the sun and d the distance between planet and sun. In Fig. 4.1 it is seen that the solar system (with the sun so much heavier than all of the planets) is in good agreement with this rule: the closer a planet is to the sun, the faster is its orbital velocity and hence the shorter is its "year." The figure includes Pluto, which because of its small size had been removed from the list of planets—today it is listed as a minor planet...

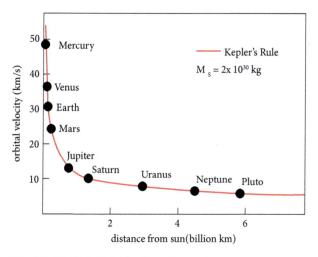

Fig. 4.1 Kepler's Rule for the solar system.

In the case of the motion of stars in galaxies, instead of planets in the solar system, the effective mass seen by a given star is that of all other stars in a sphere around the center of the galaxy and a radius determined by the distance of the star from the galactic center. The constant mass of the sun is thus replaced by an increasing total galactic mass as we move further from the center. The increase comes to an end as we reach the edge of the galaxy, and from there on, for outlying stars, we expect the $1/d$ decrease seen in the solar system—the crucial central mass now remains constant; it is the mass of all the stars in the galaxy. For the distribution of the stellar velocities we thus expect the form shown in Fig. 4.2, labeled Kepler's rule. Initially, the velocity increases as more and more stars come into play, and it then decreases as we pass the edge of the galaxy.

This, however, was not at all what one found. Instead, the orbital velocities of outlying stars remained essentially constant, even at very large distances from the center. What force held them in their respective orbits at such velocities? One could estimate the effective mass of all the stars in the galaxy through the amount of emitted light, and this mass was not nearly sufficient to account for the motion of the outlying stars. Back in 1933 the Swiss astrophysicist Fritz Zwicky therefore concluded that the actual mass of any galaxy must be some five to ten times higher than its visible mass. His idea was confirmed, largely through the work of the American astronomers Vera Rubin, Kent Ford and collaborators. Each galaxy must be embedded in a huge cloud of invisible mass, consisting of unknown *dark matter*. And the amount of dark matter in this cloud must increase linearly with the distance from the center, to give rise to the observed pattern shown in Fig. 4.2.

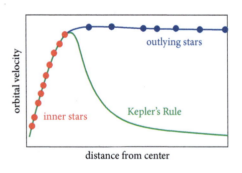

Fig. 4.2 Orbital velocities of stars in a galaxy.

So we encounter yet another enigma of gravity: not only is there no "smallest charge," but the universe contains up to ten times more matter of a completely unknown kind, subject only to gravitation. We can't see it and we have no idea from what kind of constituents it arises. Over the years, the search for dark matter particles has therefore become one of the main topics of high energy physics—so far, without success. At the European Center for Nuclear Research CERN in Geneva, Switzerland, this search is continuing in the hope of eventually identifying some new type elementary particle, subject only to gravity.

Emergent Forces

In view of these special features of gravity, it is perhaps not so surprising that since some twenty years, another point of view has appeared: perhaps the force of gravity is in fact by nature different from the others. Perhaps gravity is by nature a *collective force*: perhaps it is not a fundamental interaction between two individual elementary particles, but instead arises only such as to maximize the entropy of many-body systems. We want to follow this track in a subsequent chapter; it was pioneered some twenty years ago by the Dutch theorist Eric Verlinde.

Let us turn back to Fig. 2.3, where the opening of the divider led to gas to rush out into the previously empty section—rush out as if driven by some mystical force. Since the only result of the rush was to increase the entropy of the medium (the temperature and hence the energy of the molecules remained constant), we could speak of an *entropic force*. This force only appears for many-body systems—an individual atom does not rush out, it simply wanders around. Hence the outward drive of the gas is also often referred to as an *emergent force*—it is not present on an individual level, but emerges due to the collective effect of many.

We want to look at this in a little more mathematical detail. The first law of thermodynamics, introducing heat as a form of energy, states that the total energy E of a system decreases by the work—dW it does and increases by the amount of heat dQ put into it: $dE = -dW + dQ$. The work is in turn fixed by the pressure P exerted to change the volume, $dW = PdV$. Since the input of heat also increases the number of possible microstates, it leads to a growing entropy, with $dQ = TdS$. The resulting form of the first law, $dE = -PdV + TdS$, becomes for a one-dimensional expansion along the direction x

$$dE = -PdV + TdS = -Fdx + TdS, \qquad (4.6)$$

since the pressure is the force per area A, so that $P = F/A$, and $dV = Adx$. If the overall change of energy is zero, the constituents of the system behave as if an entropic force

$$F = T \left(dS/dx\right) \qquad (4.7)$$

is pushing them into the new empty section of the container. We should emphasize the "as if": there is really nothing pushing on any given molecule. This is a force of a different nature than the ones considered in the standard model. It only exists for many-body systems, characterized by an average kinetic energy per particle, specified by the temperature, volume and a total number of microstates, counted by the entropy. The force arises from the change of entropy, given a change of volume at constant temperature.

In conclusion: we can divide the forces found in nature into two categories: fundamental forces, acting between pairs of discrete elementary charges, and emergent forces, arising only through the collective effort of many constituents in order to increase their entropy.

What Is the Charge of Gravity?

In contrast to the numerous attempts to unify gravity with the other three fundamental forces, the Dutch physicist Eric Verlinde has pioneered the formulation of gravity as an emergent entropic force. He suggested that it does not exist intrinsically, like nuclear or electromagnetic forces, but that it emerges through the combined effort of many interacting constituents, like pressure or temperature. Matter has a microstructure, it consists of smallest fundamental units interacting through fundamental forces. We had seen that the interaction of many such constituents can lead to an additional *effective* force due to the inherent drive of the system to maximize its entropy. To show that gravity is such a force, one has to derive Newton's law of gravitation (or more generally, Einstein's general theory of relativity) from the basic laws of thermodynamics. Not surprisingly, this will require some preliminaries.

Already Max Planck had attempted to devise some smallest-dimensional unit for gravity, to make up for the absence of such a unit in nature. The mentioned attempts to force gravity into the scheme of the fundamental forces had tried to introduce a *graviton* as counterpart of the photon, but—as indicated—these attempts were so far not very successful. Planck had suggested to take the fundamental constants of nature, the speed of light c, the quantum constant h and the universal gravitational constant G, to construct a fundamental length. He thus arrived at

$$\ell^2 = \frac{Gh}{c^3} \tag{4.8}$$

for the square of what we today call the Planck length ℓ. It is in a way the smallest conceivable length, some 10^{-35} meters. To get an idea of what that means, imagine that we blow up a meter stick to a billion times the diameter of the observable universe. Only then does the Planck length increase such as to become one meter. The size of a proton, our smallest constituent of matter, is 10^{20} Planck lengths. The most powerful microscopes we have today are the high energy particle accelerators, with the Large Hadron Collider (LHC) at CERN in Geneva at the top of the list. It can resolve distances down to the miniscule scale of 10^{-25} meters; but this is still some 10 billion Planck lengths. So looking for structures of Planck length size is and will remain for quite some time outside of the range of human experiments.

Nevertheless, the Planck length does play a crucial role in our view of the cosmos. Gravity determines the large scale structure of the universe, from solar systems to galaxies. It is also the essential factor in star formation, caused by the gravitational contraction of matter clouds in space. But there is one phenomenon where gravity becomes singularly dominant: black holes. The evolution of a star runs through various stages, from gaseous clouds to ever denser stellar objects. As the star burns out through fusion radiation, the force of gravity compresses atomic matter to higher and higher densities, until finally the electrons are squeezed out and the stellar core consists of purely nuclear matter. The electric repulsion of the protons hinders further compression, but if the star is massive enough, the protons decay and only neutrons remain as constituents. The Pauli principle (no two identical nucleons in the same region of space) opposes further contraction, but if the star is massive enough (more than some 3–5 solar masses), gravity wins in the end and the dying star collapses to form a black hole, a comparatively small stellar object of immense density: a huge mass is confined to a sphere of rather small radius. A black star of ten solar masses has a radius of some 30 kilometers.

We had mentioned the determination of the escape velocity v, needed to allow a space ship to leave earth. The general form obtained for any star is

$$v^2 = 2\frac{GM}{R} \qquad (4.9)$$

with G for the universal gravitational constant, M the mass of the star, and R its radius. For stars whose mass M is large enough and whose radius R is sufficiently small, v will exceed the speed of light: $v > c$. From this point on, nothing, not even a light signal, can leave the collapsed star, which is therefore called a black hole. Its radius R has to satisfy $R \leq 2GM/c^2$. For some time it was not clear if such objects exist in our universe, but today their abundant presence in many galaxies is well confirmed, including our own, the Milky Way. Any object coming within the range of attraction of a black hole is irrevocably sucked it: it disappears below its horizon and is never seen again.

The Entropy of Black Holes

One feature of black holes is of particular interest for our issue: its entropy. The entropy of a physical many-body system is determined by the number of microstates allowed for a given macro-state, of mass M and volume V. In the case of black holes, we know only the mass and the size, nothing more. The Princeton theorist John Wheeler reported that Jacob Bekenstein had summarized this by the statement that "black holes have no hair"—we can't count any microstates, in fact, we don't even know if there are any. On the other hand, if we allow a complex object, say a volume of gas, to be sucked into the black hole, what happens to the entropy? The box of gas has a certain, even large entropy; when it disappears inside the black hole, what happens to that entropy? More generally, what is the entropy of a black hole, and how does this change when some object with a given intrinsic entropy is absorbed? Can black holes destroy entropy, in violation of the second law of thermodynamics?

The answer was provided in 1973 by the Israeli-American theorist Jacob Bekenstein, who showed that one can identify the surface area A of the black hole horizon as the entropy S_{bh} of black holes, where the unit constituent is a little area of Planckian size ℓ^2:

$$S_{bh} = \frac{A}{4\ell^2} \tag{4.10}$$

The size of black holes had already been determined by Karl Schwarzschild, in his solution of Einstein's equation of relativity: the radius of the horizon was found to be $R = 2GM/c^2$. As mentioned, the entropy generally counts the number of ways a system of many constituents can be formed; normally, that implies that it grows as the volume of the system, as we had seen above. What happens when a normal star, with its given entropy, collapses to form a black hole?

Having no information about the microstates would imply an infinite entropy, corresponding to an infinite number of possible microstates. However, Bekenstein's form imposes a limit on this: the surface area of the volume, measured in terms of basic Planck areas, represents the maximum amount of information we can have about the system. It cannot be exceeded.

If we throw something into the black hole, the absorption process must therefore cause the black hole to increase in size such that the increase in area is determined by the amount of additional entropy provided be the acquisition. It can be shown, in fact, that if a black hole absorbs a particle of mass m, its surface area and hence its entropy increase by the Compton wavelength λ ($\lambda = h/mc$) of the absorbed particle,

$$\frac{dS}{dx} = \frac{k}{\lambda} = \frac{ck}{h}m. \tag{4.11}$$

We had seen above that for systems at constant energy the first law of thermodynamics relates the entropic force to the temperature and the entropic change,

$$F = T\frac{dS}{dx} = \frac{ck}{h}mT \qquad (4.12)$$

The temperature is a measure of the average energy per degree of freedom per constituent in the medium, $T = 2E/Nk$. The total energy can be represented as the mass of the black hole, through Einstein's celebrated $E = Mc^2$, and the number of constituents is related through Bekenstein's formula to the surface area of the hole in terms of Planck areas, $N \sim A/\ell^2$. Putting all this together (see Box 5.1) gives us Newton's law of gravitation,

$$F = G\frac{mM}{R^2}, \qquad (4.13)$$

showing that one can indeed consider gravity as an emergent entropic force. The relation to black hole physics is moreover a special case of a more general situation. The holographic principle, proposed by the Dutch theorist Gerard 't Hooft and the American theorist Leonard Susskind, implies that the argumentation holds whenever an event horizon divides the world into a hidden region, from which nothing can emerge, and a region in which our laws of physics hold.

As already indicated, this suggests, though not proves, that the force of gravity is of a different nature than the three unified ones, strong, electromagnetic and weak. These three can be studied in terms of the interaction of elementary constituents, such as the attraction of a proton and an electron. For gravity, there is no such elementary charge—it sets in only for a large number of massive constituents trying to find the optimal thermodynamic state. From this point of view, the attempts to unify gravity with the forces of the standard model appear as unnatural and doomed from the beginning. The force of the air rushing out of a punctured balloon is not of the same nature as the repulsion of two magnetic north poles, and if gravity is an emergent force, it will continue to resist unification.

Before turning to the next, and quite different topic, we note a curious connection between black hole physics and the arrow of time, a connection which was implicitly used in some of the above argumentation. The surface of a black hole constitutes a horizon which can be crossed in one direction only: signals or objects can enter, but not leave. The present in time constitutes a similar horizon: we can receive signals from the past, but we cannot send any signals there. So in some temporal sense the present world is a little like a black hole immersed in the past, with which it can never communicate.

5

The Formation of Structure

And God formed man of the dust of the ground...
The Bible, Genesis 2.7

The Appearance of Order

In the world of statistical mechanics, structure means a form of order, and as such a deviation from disorder, as we had seen above. Disorder means that all possible states are equally likely. A counterpart of this definition is found in specifying information as a deviation from a random assembly of letters or bits. Randomness and disorder are the more general states, from which structure or information can arise in numerous specific ways. In this sense, the return of Joule's gas to its original volume would correspond to the monkey who randomly hits the keys of a typewriter and in this way eventually writes *Hamlet*. In both cases, the probabilities are not *exactly* zero, but immensely small...

Evidently the nature of the interaction between the constituents plays an important role in determining the state of maximum entropy, of ultimate disorder. For a system of positive and negative charges in equal numbers, the equilibrium state at high temperature and density is a hot, uniform and completely disordered plasma. At lower temperature or for dilute media, it becomes favorable for the charges to combine to form electrically neutral atoms. In that way, both charges escape from further electromagnetic interaction, and the state of maximum entropy is now a uniform, disordered gas of such neutral atoms. And if we lower the temperature still further, maximum entropy can imply a crystal, with a fixed structure and atoms oscillating around geometrically ordered positions. We see from this that under suitable conditions, the state of maximum entropy can still exhibit features that we normally associate with order, such as crystal structure.

It is for this reason that the Danish physicist Per Bak, one of the leading theorists in the formulation of what we now call complexity, asked why the Big Bang did not evolve into a hot gas of elementary constituents, particles or atoms, or into one big crystal, but instead into our present multi-faceted world. One essential part of the answer is, as we saw, the fact that the expansion of the primordial universe was simply too fast to allow the formation of overall equilibrium, of any form of maximum entropy state. Another equally important aspect is that gas, liquid or crystal as equilibrium states arise if the interaction between the constituents is weak and/or of short range. Nuclear forces are in fact extremely short-ranged, and electric forces, though in principle of long range ($\sim 1/r^2$), become effectively short-ranged, since opposite charges shield each other from the effect of others, rendering the combined system neutral. A dense plasma ball of equally many positive and negative charges does not exert any electric force on a distant charge.

The Role of Gravity

How then could our world of stars and galaxies have emerged from the primordial hot gas? The crucial aspect, we believe today, is the force of gravity. We shall return to the unique nature of gravity in a subsequent chapter; here we only note that it is of long range and there is no way to shield it. The plasma ball invisible to distant charges is clearly recognized by a distant mass: it is attracted to the ball by a force Gm_0Nm/r^2, if the ball consists of N charges of mass m each and our test mass m_0 is a distance r away; G is the universal constant of gravity. Every mass in the universe feels the gravity of all others, weakly if they are distant, but unshielded; they all add up. As a result, all large-scale structures in the universe—galaxies, stars, planets—are created by gravity; the other forces are responsible for the very small—such as nuclei—or the intermediate—such as crystals or humans.

Let us then return to the early universe. The Big Bang led to the creation of a hot gas of elementary particles, essentially in equilibrium, at maximum entropy. As a result of their formation process, there were minute fluctuations in density: some regions of space were a tiny bit higher in density then others. In a stationary system, these fluctuations would appear here and there in the course of time, and they would average out to a constant density. Our universe, however, was expanding faster than the necessary relaxation time, and this preserved the fluctuation structure and stretched these out to ever larger size. We can in fact observe all this by means of the so-called cosmic background radiation discovered in 1964 by the American astronomers Arno Penzias and Robert Wilson. They found that in whatever direction of space they looked, there was a radiation background, a steady static noise, of a constant temperature; today's measurements by space satellites give 2.725 degree kelvin with a one-per-mil accuracy.

The origin of this radiation was soon identified: it was a message from the early universe. The hot primordial plasma consisted for most of the first some hundred thousand years of charged particles (mostly protons, light nuclei and electrons) as well as photons, that is, electromagnetic radiation. Since the photons interact with the charged constituents, this early medium was opaque to light, it could not escape to radiate. We know, however, that for temperatures below about 3000 degrees kelvin, the electrically neutral atoms can start to form; for higher temperatures, the photons break them up again. The initially very hot plasma had cooled down to this "decoupling temperature" (decoupling for photons, coupling for atoms) after about 380,000 years, so from that point on, the photons could proceed unhindered. The radiation we measure today is thus the picture of the universe at this decoupling time, except that the expansion of space since then has increased its wavelength and thus lowered its temperature, down to the 2.725 degrees we measure today.

It was pointed out quite soon that the apparent uniformity of the radiation temperature should not really persist for sufficiently precise measurements. The fluctuations we had mentioned produced regions of varying density, albeit very slightly varying, and these variations should in turn cause variations in the cosmic radiation temperature. The accuracy of terrestrial measurements was not sufficient to register them, but now, with satellite data, they are very clearly seen, as shown in Fig. 5.1, taken by the Planck detector. We should remember, however, that the variations in color correspond to temperature variations of less than a thousandth of a percent, blue meaning a little cooler, red a little warmer. So we can today be sure that the primordial universe was not perfectly smooth and uniform; it contained from the beginning density variations, and these provided the seeds for the structure we see today.

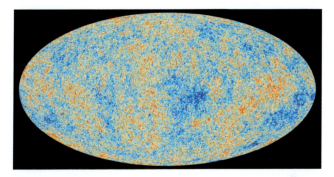

Fig. 5.1 Temperature distribution of the cosmic background radiation (from the Planck detector).

We thus have a universe consisting of randomly distributed regions of slightly varying density. In the statistical mechanics of short-range forces, this would be close to thermal equilibrium, to maximum entropy. At this point, however, gravity enters the scene; how and when is not really known, but it is now present, and a randomly distributed gas of particles is now no longer the state of maximum entropy. This is most readily seen in the framework of Einstein's theory of gravity, in which the presence of a mass deforms the space around it in proportion to its mass. In short-range statistical mechanics, we maximize entropy by distributing particles randomly over a flat surface. With gravity in force, this flat surface turns into a sheet with random depressions, which are deeper the greater the density is at that point, see Fig. 5.2. The particles which previously were randomly distributed now tend to roll into the holes and form clusters: the seeds of stars and galaxies. In the presence of gravity, the drive to maximize the entropy thus leads to cluster formation and hence to the appearance of structure.

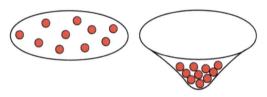

Fig. 5.2 The effect of gravity on randomly distributed particles.

It is clear that gravity can only lead to large-scale structure: the force is extremely much weaker than all other forces, so it can only play a significant role if a huge number of masses act together. The electric attraction between the proton and the electron in the simplest atom, hydrogen, is by a factor of 10^{39} stronger than their attraction by gravity at the same distance of separation; in other words, for the pairwise interaction between elementary constituents, gravity can safely be neglected. If we have a cluster of a large number N of hydrogen atoms, an outside charge feels effectively no force, since the charges in the atomic constituents of the cluster neutralize each other. The gravity forces just add up, however, so that a test particle is attracted by a composite object of mass N hydrogen atom masses. Gravity can become competitive only once there are many masses—but then it becomes very much dominant. The combined effort of many tiny forces becomes great enough to keep the earth in its orbit around the sun and the stars in the Milky Way.

The Structure of Galaxies

The almost uniform distribution of particles, still in evidence at the decoupling time, thus changed with time: the matter particles clustered together more and more, due to gravity: they rolled into the depression in space caused by the concentration of masses. Some 500 million years after the Big Bang, the universe was thus filled almost uniformly by radiation, and in this sea of photons, there was a random distribution of massive clusters, moving through space and contracting more and more with time. When two such clusters collided, they set the combined system into rotation, as shown in Fig. 5.3: this led to the disk-like structure observed today in numerous galaxies.

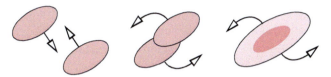

Fig. 5.3 Formation of a galaxy in the collision of two gas clusters.

So far, we have only considered the long-range effect of gravity in the formation of structure, which means that our stars don't shine. To turn on the lights, another force was needed. With increasing contraction, first the atoms and then their cores, the nucleons in the gas cluster were pressed together ever closer. The electric repulsion between the positive nucleons initially provided a resistance to further compression, but for sufficiently massive systems, this was eventually overcome by gravity, which squeezed them together more and more. This finally brought the density into the regime of nuclear forces: with sufficient pressure, four nucleons now combine to create a helium nucleus, a bound state of two protons and two neutrons, whose total mass is about 5% less than that of the sum of the individual nucleon masses. This nuclear fusion process thus liberates energy, which goes predominantly into electromagnetic radiation, photons: the star shines. The light of the stars thus arises through gravity for compression and nuclear fusion for radiation.[1]

Let us summarize the structure formation we have thus discussed. The initial state of the universe just after the Big Bang was a rather uniform hot plasma of elementary particles; small density fluctuations in this plasma, remnants of the Big Bang, under the influence of gravity subsequently formed higher density clusters, which eventually led to galaxies. The further contraction of the constituents of the clusters produced densities sufficient to overcome electromagnetic repulsion and to

---

[1] To avoid objections from the experts, we note that also the weak nuclear interaction comes onto the scene: to form the helium nucleus, two of the protons are converted into neutrons, with the emission of a positron and a neutrino in each case.

reach the nuclear fusion regime. The binding energy liberated in this fusion provides the light and thus the energy emitted by stars.

In all these transitions, the overall entropy increased. The initial hot primordial plasma was in fact in a state close to maximum entropy under the given conditions. The further expansion of the universe and the resulting evolution to a radiation gas containing matter clusters then led to media whose entropy continued to increase, but without ever reaching the maximum entropy possible. This in turn allowed an ever increasing structure formation, from dense clusters to galaxies consisting of shining stars.

Star formation is of particular interest here, since it requires three different forces of nature. Gravity is crucial to contract the cluster of matter to ever higher densities, and to continue doing this even against the resistance of the electric repulsion of the protons in the cluster. Finally the gravitational contraction brings the cluster to a density high enough for nuclear fusion to set in, nucleons combine to nuclei, liberating energy which is emitted as radiation of light. Let us see what this means for the entropy evolution. The entropy of the matter cluster before reaching fusion density is certainly higher than that of the star resulting from the further contraction; but the radiation emitted in the nuclear fusion is of so much higher entropy that the entropy of the shining star, matter plus light, is many orders of magnitude higher than that of the star before it started to shine. Since it continues to shine and since further stars are still being formed, equilibrium and hence maximal entropy has not yet been reached. Both entropy and order are increasing.

6

The Energy of Space

Facets of Energy Conservation

The conservation of energy is one of the most fundamental axioms for physical processes, and it leads to several different forms of energy. The flexed bow held by an archer holds considerable potential energy, and when the archer releases the arrow, this energy is not lost, but is converted into kinetic energy of the flying arrow. When the arrow hits its target, the kinetic energy is converted into heat, formed by the impact. Until the advent of thermodynamics and the realization that heat is a form of energy, it was thought that stopping or slowing down motion by friction meant a loss of energy. The lesson to be learned from this is that when we encounter an apparent loss or gain of energy—a seeming violation of energy conservation—we should consider the possibility that it is simply converted into or arises from a new form of energy.

A striking instance of such a generalization came with the advent of general relativity, proposed in 1915 by Albert Einstein, relating the structure of the universe to the matter and energy it contains, the curvature of space to the gravitational forces of the matter in that space. In the words of the Princeton theorist, John Wheeler, "matter tells space-time how to curve, space-time tells matter how to move." It thus becomes evident that one cannot expect a conservation of energy independently of the behavior of the space-time it is contained in. We cannot put matter and radiation into a flat static background space and expect that space to remain that way.

The Advent of the Big Bang

In mathematical terms, the introduction of matter or more generally of energy-momentum M into an initially flat background space results in curvature, measured by the Ricci-tensor R,

$$-R = 8\pi GM; \tag{6.1}$$

here G denotes the universal gravitational constant. For the time (t) evolution of the spatial length scale $a(t)$, the Einstein equations led to

$$(da/dt)^2 = 8\pi GM/3a \tag{6.2}$$

where M denotes all matter (light and dark) and electromagnetic radiation in the universe, providing a relation between the expansion of space and its content. This equation was soon obtained and solved independently by the Russian theorist Alexander Friedmann and the Belgian physicist and Catholic priest Georges Lemaitre. It is now known as the Friedmann-Lemaître-Robertson-Walker form and leads to a space expanding in time,

$$a \sim M^{1/2}t^{2/3}, (da/dt) \sim M^{1/2}t^{-1/3} \tag{6.3}$$

with a slowly decreasing expansion speed. From this, Lemaître concluded that everything must have started from a tiny spot of immense energy which subsequently blew up to create our present universe: the idea of the Big Bang was born. The entire picture seemed in fact quite reasonable: the initial "explosion" caused an expanding bubble, and as the initial energy was being used up, the expansion slowed down.

Einstein strongly disagreed with this view and held onto the static universe; he wrote to Lemaître that while his mathematics was correct, his physics was "abominable." Here we should, for the moment, suppress our hindsight knowledge and recall that Einstein's point of view, and that of almost all his colleagues, was in fact at that time quite reasonable. He assumed that you could start with given static space-time and then describe the behavior of matter and radiation within this fixed world. The idea that putting a chair into a room (even a galactic chair into an extra-galactic room) would make that room expand—well, that just did not sound reasonable.

To resolve the "problem," Einstein introduced what is now called the "cosmological constant," Λ. Effectively this meant adding some fictitious unknown energy of empty space to the matter and radiation present, so that its gravitation would reduce the expansion. For an infinite flat background space and with a constant vacuum energy density Λ that leads to

$$(da/dt)^2 = (8\pi GM + \Lambda a^3)/3a. \qquad (6.4)$$

This does not take care of the expansion; in fact, we shall see that it does the opposite, it enhances it. Einstein therefore formulated the problem in a closed curved space. This changes the equation (6.4) to

$$(da/dt)^2 + 1 = (8\pi GM + \Lambda a^3)/3a, \qquad (6.5)$$

and now it works: this equation indeed allows a static finite universe, with $(da/dt) = 0$, $a^2 = 1/\Lambda$ and $\Lambda = (4\pi GM)^{-2}$. Einstein's world thus was closed, finite and static, as he had wanted. All this turned out to be very short-lived, however.

The scenario was destroyed by two decisive blows, both in the 20th century. In 1929, the American astronomer Edwin Hubble showed in a study of distant galaxies that the observable universe is in fact expanding, and in 1998 two teams of American and Australian astronomers showed through a study of super-nova explosions that this expansion is in fact accelerating. Although Hubble's discovery ranks undoubtedly among the most important ever made in natural science, he was never awarded the Nobel prize. In the following years, the responsible Nobel committee changed its attitude towards astronomy, and in 2011, the leaders of the super-nova experiments, Saul Perlmutter and Adam Riess from the USA and Brian Schmidt from Australia, did receive this prize. We will now look in more detail at the expanding universe.

Horizons

The horizon is the limit of our field of vision. As such it has been essential for our view of the world since antiquity. The hull of a ship sailing out to sea disappeared from sight, beyond the horizon, before its masts did, telling us that the earth is round, not flat. Yet we can never reach the horizon—if we try to approach it, it fades away before us, just as we can never reach the end of the rainbow.

If we would have looked up into the night sky in pre-relativistic times, we would have expected to see an infinite universe. Subsequently, relativity theory taught us that the speed of information transfer, the speed of light c, is finite, so we could only see those objects whose light has had enough time to reach us. That's why the night sky is black for us—we can only see the stars that are close enough. If a far-away light source, a distance d away, is turned on now, we will see the light only at a time $t = d/c$ later. The further away it is, the longer we have to wait, for a time span determined by the speed of light. The sunlight arriving on Earth at any given time was emitted by the sun eight minutes earlier... What happens if we consider much more remote cosmic objects?

To describe scales in the universe, astronomers find it useful to measure distances in lightyears (ly), that is, in units corresponding to the distance that light travels in one year:1 $ly = 3.8 \times 10^{12}$km. The sun is then only 3.9×10^{-5} ly away from us, and our own galaxy, the Milky Way, has a radius R_{MW} of about 53,000 ly. When cosmologists talk about the large-scale structure of the universe, they are thinking of extra-galactic distances, much larger than R_{MW}, and in this regime, the past century brought the two mentioned dramatic discoveries. Both completely changed our view of the universe.

Edwin Hubble was working at the Mount Wilson observatory near Pasadena in California, which featured the most powerful telescope available at the time. He had found that the shining distant nebulae seen beyond the milky way at distances much larger than R_{MW} were in fact galaxies on their own. And when he investigated their behavior, he found something totally unexpected: with time, these distant galaxies moved away from us. This immediately led to a rebirth of the original Einstein formulation.

In the pre-Hubble scenario it was assumed that space holds still in the temporal evolution of the universe, and precisely that was ruled out by Einstein's original equations. They showed instead that the dimensional scale of space increased with time in a form determined by gravity. A simple illustration of the result of this phenomenon is given by an ant crawling on the surface of a balloon (Fig. 6.1), maintaining a fixed "ant" speed. Starting from the equator of the balloon, the ant would thus need a given time to reach the North Pole. However, if during the trip the balloon were to be inflated to ever larger size, this time would become longer and longer, and if the balloon was inflated rapidly enough, the poor ant would never reach the pole. In fact, although it kept crawling, the pole would move further and further away.

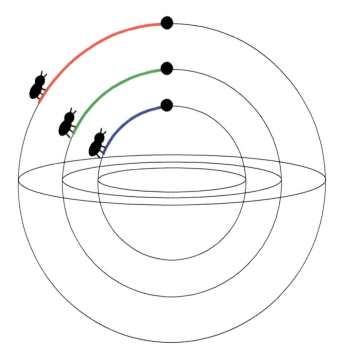

Fig. 6.1 The ant on the balloon

Thus the signal of light reaching us from a far-away cosmic source has been moving towards us with its local speed of light, but the distance covered increased with the expanding universe. As the light was underway, the size of the universe increased, moving the light position at any given time, as well as its original source position, further away from us. And if the universe grew fast enough, exceeding the speed of light, the signal would never reach us. The corresponding horizon of communication thus depends on the expansion rate of the universe as well as on the local universal speed of light. Here we should note that an expansion of space faster than the speed of light is not in conflict with relativity theory: that only limits the speed of causal signals, of information transfer, and an expanding universe cannot send a signal from one point to another.

Hubble's Law

What Hubble had found was that from our point of view, far-away objects (galaxies in his case) move out from us, recess, in all directions. The speed of the object in question was determined by the red-shift of the light it emitted. This effect is similar to the Doppler shift of sound sent out by a rapidly moving source, such as a train whistle: its frequency is shifted to lower values the faster it moves away from the listener. So one could order the observed distant light sources according to their recession speeds, and it was evident that our universe is in fact expanding, in accord with the solutions which Friedmann and Lemaître had obtained for Einstein's relativity equations.

To specify the exact form of the expansion, one also had to know the distances of the galaxies from us. This was achieved by measuring the brightness of so-called Cepheid stars, which were pulsating with a well-defined period and amplitude. The variation of these observables effectively provided a metric scale for the universe and hence allowed distance measurements for the receding galaxies. Using these, Hubble found that the more distant galaxies recess faster than the closer ones, as illustrated schematically in Fig. 6.2. Today we can use the brightness of super-nova explosions as a more precise tool for distance determinations.

There was only one way to account for the observed pattern: a universal local speed plus an expansion speed depending on the distance from the observer. Here again the picture of the expanding balloon is very appropriate: as the balloon expands, any two points on it, wherever they

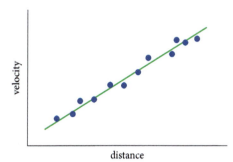

Fig. 6.2 Recession speeds vs. distances for remote galaxies (Hubble)

might be, separate in the same way; our position is in no way special. And the further away the galaxy is, the greater is its expansion-induced speed with respect to a given observer. The result is Hubble's law

$$v = H_0\, d, \tag{6.6}$$

where v is the recession velocity of the object in question and d its distance from the observer. The slope of the increase in Fig. 6.2 thus defines the Hubble constant H_0 in units of inverse time.

Since v gives the velocity seen by us, distance per time, and d the distance seen by us, their ratio

$$\tau_0 = \frac{d}{v} = \frac{1}{H_0}, \tag{6.7}$$

determines the age of the visible universe. Present measurements, more accurate than those of Hubble, give $\tau_0 \cong 14$ billion years. In a static universe, that would imply about 14 billion lightyears for the radius of the observable universe. However, the expansion of the universe during the time the light travels greatly increases the present distance between observer and original light source. Thus by the time the light from the primordial universe reaches us, it has travelled a distance of 44 billion lightyears, with an effective speed of more than three times the universal causal speed of light. For us, the diameter of the universe from which we have received information is now close to 90 billion lightyears.

After Hubble's discovery, Einstein realized that he had missed the chance to predict an expanding universe, and he called the cosmological constant, Λ, his biggest blunder. As it turned out, he was wrong also on this point. The experiments of the super-nova study groups could not only determine the recession velocity of distant objects, they could also show that this recession velocity in fact increases with time—the galaxies move outwards faster and faster. Einstein's equations, however, predicted in their original form without cosmological constant (see equation (6.3)) that the velocity should decrease with time. There was only one way to change that: a cosmological constant Λ. In Einstein's original proposal in curved space, it was there to prevent the expansion; now we want to speed it up. And that is in fact what happens if we introduce the cosmological constant in flat space, as shown in equation (6.4). The mysterious medium filling all of space with a constant energy density is back; this medium is now called:

Dark Energy

In our world, in the solar system or even in the Milky Way, you cannot notice it in any way, see it or feel it; its effect comes into play only on extra-galactic scales. As mentioned, the presence of the cosmological constant means that the vacuum is filled by a medium of low but constant density, a density that remains constant as space expands. The Friedmann-Lemaître-Robertson-Walker counterpart of equation (7.4) for the acceleration is given by

$$a^2(d^2a/dt^2) + 4\pi GM = \Lambda a^3/3 \qquad (6.8)$$

and with the measured values for the recession acceleration, it can be used to determine the density Λ of dark energy in space. As already mentioned, the resulting density is extremely low, equivalent to some few protons per cubic meter, so it has an effect only over cosmological distances—within our galaxy, its impact is negligible. Nevertheless, on the whole its contribution is dramatic. Using equation (6.4), we can determine the combined contribution of mass (light and dark), radiation and dark energy to the overall energy input. The visible matter and radiation in the universe can be measured, and it amounts to some 5% of the whole; the additional dark matter gives another 25%. We thus have to conclude that dark energy, not noticeable on our local scales, provides 70% of all the energy content of the universe. What we can see, describe and understand after two thousand years of natural science is only 5% of all there is…

As a result, we are now faced by two crucial questions: what is this medium we call dark energy? And if its density remains constant under expansion, if M/V remains constant as V grows, where does the increase in invisible mass and energy M come from?

Our present answer to the first question can be simple: we have no idea. Any known forms of matter have been ruled out. We just don't know what it is. For a while, theorists thought they could solve the puzzle. The result is now called the vacuum catastrophe. Quantum field theory pictures the vacuum as a sort of quiet water, with fish swimming below the surfaces. These fish can briefly emerge above the surface, but only for an infinitesimally short time, then they fall back into the sea. Such fluctuations lead to a fluctuating vacuum energy, but the corresponding calculations led to a value which is a factor of 10^{129} too high, compared to the measure value of Λ. With some delight, this was claimed as the "wrongest" result ever reached in physics.

The other question is "under discussion"—the scientific form of registering disagreeing opinions. The starting point is the mentioned statement of John Wheeler: "matter tells space-time how to curve, its curvature tells matter how to move." A generalized version might be "matter tells space how to expand, the expansion of space creates the needed matter." So we have to consider how energy conservation might take place in an expanding universe. In classical mechanics, the overall energy is not only conserved; its actual value is also open up to a constant, to be defined by the observer: for $E' = E + const.$, we have $(dE'/dt) = (dE/dt) = 0$. In general relativity, this freedom is removed: adding a constant to the overall mass and radiation changes the spatial behavior, see equations (6.1) and (6.2). This is sometimes expressed as "there is no energy conservation in general relativity."

As a preliminary exercise, we consider the cosmic background radiation discussed in Chapter 5. It was freed some 300,000 years after the

Big Bang, when the formation of electrically neutral atoms made the universe transparent to light, so from then on, the primeval photons moved freely through space. At that starting time, the photon radiation had a temperature of some 3000 degrees kelvin, the binding limit for atoms. The radiation we receive today has a temperature of about 3 degrees kelvin; more precisely, it has 2.725 ± 0.001 degrees kelvin, in whatever direction we look. This decrease corresponds to the cosmic redshift: the expansion of space causes a stretching the wavelengths of the radiation. The number of cosmic photons has remained constant since their liberation—they did not interact afterwards. So the energy of the cosmic radiation, measured by the temperature, has decreased by a factor of 1000 since their emission. The energy of the sum of matter and radiation is evidently not conserved. What happened to all that lost radiation energy?

In the spirit of equation (6.3), we could say that it was used to reduce the expansion speed a little: it plays the role of friction in slowing down the expansion of space. In other words, if we reduce the right-hand side of that equation a little (the radiation is contained in M), then as a result the left-hand side is reduced correspondingly, and space now expands a little slower. So if we want to have some form of energy conservation, we have to take that into account: it is the sum

$$(da/dt)^2 - 8\pi GM/3a = 0 \qquad (6.9)$$

that is conserved in the process. It has therefore been proposed (see the cited papers of Philip Gibbs) to consider

$$E = M - \frac{3a}{8\pi G}\left(\frac{da}{dt}\right)^2 = 0 \qquad (6.10)$$

as the total energy in an expanding spatial volume $V \sim a(t)^3$. It is not only conserved, but according to the Einstein equations, it also has to vanish, removing the renormalization freedom of classical mechanics.

A similar effect in daily life is given by a treadmill as used for exercise: by running, you move the elastic band you stand on. The faster you run, the faster the band moves, and the greater is the effective distance covered by that moving floor. And as you get tired, you run slower and hence the distance becomes smaller. To achieve a more complete parallel with the microwave radiation, we would have to take a treadmill already moving with a certain intrinsic speed, which you can then enhance by running, or recover by slowing down.

Energy input and space motion output are thus intimately connected, and one cannot simply talk about energy conservation. The energy of the microwave radiation is an intrinsic factor for the space expansion, and if it is reduced, so is the expansion. What is conserved is the sum of energy input and motion output.

We now return to the dark energy and rewrite equation (6.10) accordingly as

$$E = M + \frac{\Lambda a^3}{8\pi G} - \frac{3a}{8\pi G}\left(\frac{da}{dt}\right)^2 = 0. \qquad (6.11)$$

The first term on the right-hand side contains mass (light and dark) and radiation, the second the dark energy, which joins forces with M to increase the energy input; the third term compensates the effect of these two by providing an expanding space. Since this term can be arbitrarily large in absolute value, so can the dark energy contribution Λ. Equations (6.8) and (6.11) thus correctly describe the universe as it is observed now, and as long as the dark energy density Λ remains a constant, does not wear down, the universe will continue in its accelerating expansion.

Equation (6.11) illustrates the problem with energy conservation in general relativity: if space were static, which according to the Einstein equations it is not, or if we find some other reason to define only the first two terms as energy, then we would claim non-conservation of energy. If, on the other hand, we include the potential stored in the spatial third term, then the overall energy is conserved and vanishes.

A Universe from Nothing

Another often noted puzzle is the constant energy density Λ of "empty" space. From equation (7.11) we conclude that the "objects" in space, giving rise to an energy, are matter (light and dark) and radiation, contained in M, plus dark energy, contained in Λ. As the universe expands, the dark energy density Λ remains constant. As already mentioned, this naturally raises the question: where does this energy come from? In a more general sense, this question already arises with the Big Bang: where did the energy of the nascent universe come from?

An interesting, though not universally accepted answer was first introduced by a young assistant professor at Columbia University in New York. He proposed that the Big Bang, the appearance of the universe, was simply a vacuum fluctuation, which subsequently expanded. In Tryon's own words, "In answer to the question of why it happened, I offer the modest proposal that our Universe is simply one of those things which happen from time to time." This did not imply a violation of energy conservation in the wider sense of general relativity: the increase of the combined mass and dark energy was compensated by the geometric gravitational curvature term, the third term in equation (6.11). This equation implies that our traditional energy conservation only holds at sub-galactic scales, when the effect of geometric space curvature is negligible. At extra-galactic scales an effective gravitational potential appears, in the from of space curvature, and this potential exactly cancels the energy of matter, radiation and dark energy. The Big Bang then did not require an introduction of energy from the outside, in order to take place. It was "just one of those events" and it could regulate its energy balance on its own.

Further Reading

The idea of a universe from nothing was introduced in
Edward P. Tryon, *Is the Universe a Vacuum Fluctuation?*, Nature 248 (1973) 396.

It is further elaborated in
Victor J. Stenger, *The Universe: the ultimate free lunch*, Eur. J. Phys. 11 (1990) 235

and in much detail in
Lawrence M. Krauss, *A Universe from Nothing*, Free Press, Simon and Schuster, New York 2012.

7

Critical Behavior

Great flees have little fleas
upon their backs to bite 'em.
And little fleas have lesser fleas,
and so ad infinitum.
And the great fleas themselves, in turn,
have greater fleas to go on,
while these again have greater still,
and greater still, and so on.

Augustus de Morgan (1801–1871)
(paraphrasing Jonathan Swift)

Freezing and Melting

Critical behavior is a phenomenon which can only occur in complex
systems consisting of many individual constituents. A single particle or
two cannot freeze or melt. Let us look at what constitutes such behavior.
We had already noted that in order to define *critical* behavior, we first
have to define *normal*, that is, non-critical behavior. In physics, *normal*
generally means that small causes have small effects. If we heat water a
little, it usually gets a little warmer, and if we cool it, it gets a little colder.
That's what we consider to be normal behavior. But if we heat it a little
when it is already at 99 degrees Celsius, it boils and turns into steam.
And similarly, if we cool it a little, starting at one degree Celsius, it becomes ice. So while normally a temperature change of one or two degrees
has little effect, there are *critical* points, zero and one hundred degrees,
where such a small change produces a dramatic change in the state of

the medium. Transitions of this kind, from water to ice or from water to vapor, are what we call critical behavior.

In nature, one finds numerous such transitions. Metals melt at a specific temperature. Iron becomes magnetic below a definite temperature, the Curie point, named after the French physicist Pierre Curie. Mercury is a normal metal, with resistance to current flow, down to a certain critical temperature; below that, it loses all resistance and become a superconductor. Similarly, helium below a specific temperature contains a component of liquid which flows in a non-viscous fashion, makes an ideal superfluid. The very early universe, up to a critical hadronization point, was an immensely hot and dense plasma of quarks, without any empty space in between; that only appeared after it had cooled down below the critical hadronization temperature. So our present world, with objects distributed in an empty void, first appeared at a critical point, was created in a transition.

Singular Behavior

The sudden change implied by critical behavior led physicists, for a long time, to put these phenomena aside—they were not normal. Similarly, mathematicians preferred to deal with *analytic* functions, which change smoothly as some control variable is changed. At critical points, that was no longer the case; the relevant functions, such as pressure or density, here become *singular*: they change abruptly. So for normal behavior, one could readily formulate theories, such as the ideal gas law, which specifies how pressure changes with volume and/or temperature, and one could derive these laws from a mechanical substructure of the medium, from molecules flying through space. At a critical point, such formulations break down, and for a long time, one only had certain recipes to describe the happening. If a specific observable $O(x)$ becomes singular when a control variable x approached zero, this was formulated as $O(x) = 1/x^\alpha$, defining a *critical exponent* α. One had studied a variety of such observables, such as specific heat, compressibility, magnetization, susceptibility and more. For each, a critical exponent was defined and named, leading to a set of critical exponents labelled by Greek letters, α, β, γ, δ etc. To illustrate, the specific heat $C_v(T)$ generally diverges as the temperature

T approaches the critical point T_c, and this was then formulated as

$$C_v(T) \sim |T - T_c|^{-\alpha}, \tag{7.1}$$

where $|x|$ implies the absolute value of x. The critical behavior of a given system was thus essentially identified and catalogued through the set of critical exponents. Through thermodynamic arguments, it was possible to formulate various relations between the different exponents, relations usually named after their discoverers, such as the Rushbrooke inequality,

$$\alpha + 2\beta + \gamma \geq 2, \tag{7.2}$$

the Griffith inequality, the Widom inequality, the Fisher inequality and more. This encyclopedic collection of exponents and their relations was the state of the art until the 1970s—in other words, up to a time at which relativity theory and quantum theory had been well-established as formalisms in physics for half a century. It required new ideas to bring the critical behavior of complex systems under control, and it is perhaps too early even today to claim that this is fully achieved.

The basic ideas for the new "critical" physics were proposed by two American theorists, Leo Kadanoff of the University of Chicago and Kenneth Wilson of Cornell University. Since also statistical mechanics in general is based to a considerable extent on the work of Josiah Willard Gibbs of Yale University, we are here dealing with a field of physics whose foundations are remarkably American... although first ideas do go back to Galileo and even earlier.

Scale Invariance

Normally, the appearance of what we see depends on the scale at which we look. Seen from an airplane, a forest looks like a rather uniform green surface. As we zoom in, it acquires structure of more and less dense growth, until still closer we discern individual trees. And these turn out to consist of branches and leaves. As seen from outer space, the surface of the Earth consists of multi-colored patches, which become oceans, deserts, mountain ranges and more as we reduce our scale of vision. So there are different scales, and at each scale, we have a characteristic measure. But there exist structures which look the same no matter what our scale of looking is, which become *scale-invariant*. Already Galileo had noted in his celebrated work on "Two New Sciences": *I do not see that the properties of circles, triangles, cylinders, cones and other solid figures will change with their size.* Mathematicians subsequently constructed more complex figures which are *self-similar*: they consist of elements which are of the same nature no matter how detailed our view is. The French mathematician Benoit Mandelbrot developed the study of fractal behavior as a new field of mathematics, in which such structures are investigated. A celebrated example is the so-called Sierpiński triangle, introduced 1915 by the Polish mathematician Wacław Sierpiński (Fig. 7.1).

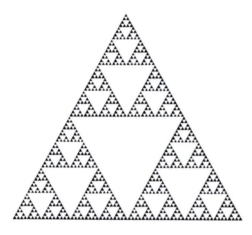

Fig. 7.1 The Sierpinski triangle.

Another often cited example is the coastline of Norway—whatever scale we use to look, it appears jagged, and if we are given a black and white picture, we cannot tell at what scale it was taken. We shall return for a closer look at fractal behavior in a subsequent chapter.

The Ising Model

With these ideas in mind, Leo Kadanoff approached the behavior of spin systems, which provide the basis for the study of magnetism. To keep things as simple as possible (but not simpler, to follow Albert Einstein),

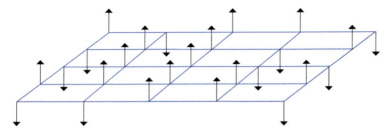

Fig. 7.2 The Ising model.

we imagine a large two-dimensional grid, on which at each intersection point there is a spin, a vector arrow of unit length, pointing either up or down; see Fig. 7.2. Each spin interacts with its nearest neighbors only, and in such a way that aligned spins (pointing in the same direction) are energetically preferred to nonaligned. This becomes an issue, since the whole system is held at a specific temperature, and the presence of thermal energy causes the spins to flip up and down randomly in time. Interaction wanting alignment and thermal motion flipping the spins around thus compete with each other. This model is known as the Ising model, named after the German physicist Ernst Ising, who studied it in his 1925 doctoral thesis, without, however, arriving at a solution to the question of who would win in the competition. That was achieved only twenty years later by the Norwegian-American theorist Lars Onsager, in what remains, even today, a veritable mathematical tour de force. It provides yet another illustration of the difficulties that complex systems create in theoretical physics. We had noted that this is about as simple a model as one can think of; yet even now, after various attempts, can it be solved analytically only in the case of a two-dimensional grid, and with great effort, as mentioned. For the three-dimensional case, there is still no analytic solution; we only have numerical calculations. We have arrived at the limits of our present mathematics.

However, the resulting solutions of the Ising model, exact where possible, approximate or numerical otherwise, are very instructive. At high temperatures, the thermal energy causes the spins to rapidly and randomly flip up and down, overcoming any attempts to align them. If we average over all spin orientations at a given time, we thus get essentially zero: as many point up as down. When the temperature is reduced, the interaction driving the spins towards alignment becomes more and more effective, creating clusters, islands of aligned spins in a sea of randomly aligned ones. And at a certain temperature T_c, at the counterpart of the Curie temperature for this model, we get a cluster of aligned spins spanning the entire grid from one side to the other. There still remain different sized smaller clusters of random spins, but the big one makes sure that the average over all spins now is no longer zero. It can be more up, or more down, the chances are the same, but it has to be one or the other. The phenomenon is in physics referred to as *spontaneous symmetry breaking*: from high temperatures down to T_c, the average spin is zero, so that flipping all spins to their opposite orientation changes nothing, leaves the state of the system invariant. Below T_c, that is no longer the case: now it's either one or the other for the average spin. The only remainder of the symmetry is that each is as likely as the other. At the transition, the symmetry of the state is thus *spontaneously* broken, without the action of any agent. As the temperature is decreased further and then approaches the absolute zero point, all spins are aligned with each other, and the average becomes unity. In Fig. 7.3 we show the behavior of the average spin value, the magnetization $m(T)$, as a function of the temperature.

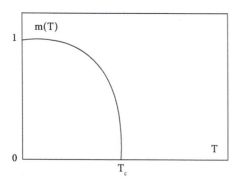

Fig. 7.3 The magnetization $m(T)$ in the Ising model, as function of temperature T.

Incidentally, the characterization of the transition in terms of symmetry breaking also provides an explanation for the abruptness of the change at the critical point. Symmetry breaking is a "yes-no" phenomenon: you cannot break a symmetry "a little," just as you cannot die a little. We had seen a similar situation in our look at percolation: there is a global connection, or there is not. And in fact the critical point in the Ising model can also be obtained as the point for which suitably defined clusters span the entire system.

Renormalization

Kadanoff now concentrated on the world of many clusters very close to the transition: it is not so different from the coastline of Norway, jagged no matter what measuring stick you use. So he divided the grid into blocks of say nine spins and defined for each block a new spin fixed by the majority spin value of its members, either up or down. We now have a new grid, with such newly defined spins, but it is of the same structure as the old. We continue the procedure, blocking again and again. This leads to three different limits, depending on the temperature; the physicists speak of *fixed points*. At very low temperatures, for $T \to 0$, the result of many blockings is a structure of all spins pointing up, or all down: we obtain a fully ordered state. For the other extreme, with $T_c \to \infty$, the blocking converges to a completely random arrangement, with vanishing average spin. But there is a third solution, a critical temperature T_c; as we approach T_c, there are aligned spin clusters of all sizes, and the picture we get remains the same for successive blockings. At the critical point, the state of the system has become scale-invariant: the structure is the same, no matter what scale we use. Though the pattern is less orderly, we have here also the self-similarity of the Sierpiński triangle: subsections of subsections still look like the starting section.

Starting from the observation of scale invariance around the critical point, Ken Wilson developed what today is known as renormalization group theory. In the normal world, we define scales in terms of some basic observables, such as the size of an atom or the length of the equator. And in our spin system, we have at each temperature above T_c a typical cluster size—in statistical physics referred to as *correlation length* λ, and this provides the relevant scale of the system. At the critical point, however, we have domains, clusters, of all sizes, from the smallest to the largest, the correlation length becomes infinite. Now we have no more scale: the system at T_c has become scale-invariant. We can define our blocks however we want to: the physics structure of the resulting system is the same.

Mathematically, a change of scale is denoted by $r \to r' = br$, turning the length r into a length br, one meter into ten meters, for example. We want the physics to remain unchanged under this shift; let us see what that means. The spin system of the Ising model is characterized by two basic control variables: its temperature T, determined by the outside energy put into the system, and a possible external magnetic field H applied to it. So far, we had not considered the latter, we had assumed that there is no outside magnetic field. If there were, it would force the individual spins to align in its direction, thus making the average over all spins different from zero at any temperature, also above T_c. We thus have two possible control inputs: energy, leading to temperature, and an outside magnetic field, leading to spin alignment. Whatever observables can be measured for the system, they will then depend on these two parameters T and H. So when we now change the scale of r by b, we have to change T and H accordingly, to ensure that all relations remain the same. This means that when

$$r \to r' = br, \text{ we get } T \to T' = b^x T \text{ and } H \to H' = b^y H,$$

in terms of two critical exponents x and y. Instead of the collection of up six exponents for a given system, specified by the behavior of various observables, we now have just two, specifying the scale behavior of the two control parameters temperature T and external field H. Through the required scale invariance of the observables, all the previous exponents α, β, γ, ... are now determined in terms of the above exponents x and y; for example, we find it for the two-dimensional Ising model $\beta = (2 - y)/x$. Wilson's underlying theory is generally referred to as *renormalization group theory*, since the scale transformations carried out by a shift b of the length scale have the property that two successive shifts are again a shift—leading to a structure mathematicians refer to as a *group*.

To obtain some idea about the progress brought about by the developments of the scaling theory of Kadanoff and Wilson, let us recall the state of astronomy before Kepler and Newton. In the prior world of Ptolemy, planets moved on epicycles superimposed on cycles around the stationary earth. These had to be calculated for each planet separately, involving a number of empirically determined parameters. Using quite involved mathematics, it was possible to specify planetary positions quite accurately—but it was more of a collection of recipes, not really a theory. That finally came when Newton formulated the law of gravity, involving only the masses of the attracting bodies and one universal constant— valid everywhere, from falling apples to the lunar motion around the earth to stellar systems in distant galaxies. In some way, the collection of critical exponents and their inequalities is Ptolemaic in nature. In Wilson's renormalization group theory, each system leads to two basic critical exponents, corresponding to the two control parameters temperature and external field. Using these exponents, all the others could be determined, as well as the inequality relations between them. And this formalism holds quite generally for critical behavior—for liquids, magnets, quark matter and more.

Strong Interaction Thermodynamics

Let us stay with quark matter for a while. With that name, or more specifically, with the concept of the quark-gluon plasma, we refer to matter at extreme temperature and density, such as was existing in the first ten microseconds of the universe, just after the Big Bang. It is matter interacting through the strong nuclear force, the same force that binds nuclei. In 1973, the fundamental theory for this form of interaction, *quantum chromodynamics (QCD)*, appeared, with quarks as the basic constituents and gluons as the force carriers. It was proposed by Harald Fritzsch, Murray Gell-Man and Heinrich Leutwyler, working at the California Institute of Technology in Pasadena, California. Thus it seemed in principle natural to construct the statistical counterpart of quantum chromodynamics to study strongly interacting matter, just as one uses mechanics to formulate statistical mechanics. The problem was the *strong* aspect of strong nuclear forces: the usual methods of solving quantum field theories, with quantum electrodynamics as prototype, assumed that the interaction only produces relatively weak perturbations of a non-interacting system. This perturbative approach fails totally for the most interesting aspects of quantum chromodynamics, in particular in the temperature/density regime in which the transition from unbound quarks and gluons to the hadrons (protons, neutrons and mesons) of our present world takes place. A new approach was needed for this critical region.

It was proposed in 1974 by Ken Wilson on the basis of his renormalization theory: one could construct a model of QCD on a lattice grid, similar to the Ising model mentioned above, but now making both space and time discrete. The actual theory would then be recovered by letting the lattice spacing go to zero. The resulting model is indeed not so different from spin models, for which one had computer methods of solution. In 1979, a group of theorists at Brookhaven National Laboratory near New York (Michael Creutz, Laurence Jacobs and Claudio Rebbi) formulated dedicated computer techniques to address the problems, and this introduced *computer simulation* as the needed new approach to QCD. The idea is both ingenious and straightforward. One constructs on the computer a lattice and builds on this lattice a system, a world as prescribed by QCD. Given this world, one can now carry out the measurements needed to obtain the desired observables. In the limit of small lattice spacing and given enough measurements, these computer observables then become identical to those in the "real" world.

By 1981, three groups had independently applied computer simulation to the study of strong interaction thermodynamics, one in Budapest, one at MIT in Boston, and our group in Bielefeld. The results showed that by increasing the temperature, one indeed moved from a region in which quarks were confined to exist as hadrons to one in which this confinement was ended. Quantum chromodynamics thus predicts as expected that strongly interacting matter experiences critical behavior, in a phase transition from normal hadronic matter to a plasma of deconfined quarks.

And again the result is of quite universal nature. The role of the coins in percolation or the aligned spins in the Ising model is now played by the hadrons. As long as they are well separated, we have hadronic matter, in which the quarks are combined to form hadrons. When they overlap sufficiently, at high density, the quark constituents of the hadrons cannot identify a specific parent—one just has a large system of interacting quarks, quark matter. In Fig. 7.4, we illustrate the transition from nucleon matter to the corresponding quark medium. From this point of view, the onset of quark matter corresponds to the percolation of hadrons, just as the onset of magnetization arises from the percolation of clusters of aligned spins.

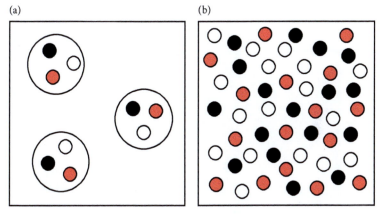

Fig. 7.4 From nucleon matter at low density (a) to quark matter at high density (b).

In 1986, an international conference was held at Brookhaven National Laboratory, dedicated to the status of the investigation of critical behavior in quantum chromodynamics. The photo, Fig. 7.5, shows some of those involved. The interest of Brookhaven National Laboratory in the field was not quite academic: in 1986, plans had been set in motion for the construction of a large nuclear accelerator, with which strongly interacting matter and the quark liberation transition was to be studied experimentally. The mentioned conference was one of the preliminary steps towards this aim, and in the year 2000, the accelerator was completed and went into operation. Since then, it has provided numerous pieces of evidence for a new state of matter in QCD.

The ultimate experimental discovery of the quark-gluon plasma (QGP)—that is, of a thermal medium consisting of unbound quarks and

Fig. 7.5 The 1986 Conference on Lattice QCD at Brookhaven National Laboratory, with Claudio Rebbi and Ken Wilson (sitting, left to right) and (standing, left to right) Michael Creutz, Sid Kahana (leader of nuclear theory at BNL) and the author.

gluons—has in fact turned out to be as much of a political as a scientific issue. Large amounts of funding were invested in Europe (CERN, Geneva) as well as in the USA (Brookhaven, New York), with the aim of finding the QGP. There were various indicators for QGP formation, but in each case, an alternative explanation could not be excluded.

To satisfy the funding agencies, at several times discovery claims were made public, with partial supportive evidence. As of today, it has become clear that high energy nuclear collisions do indeed produce media of a higher energy density than ever before studied in the laboratory, and that these media show features that can only be understood in terms of quark-gluon constituents. On the other hand, it is becoming more and more evident that these media are non-equilibrium systems and as such cannot be directly compared to the equilibrium QGP studied in lattice QCD. So it evidently became more and more necessary to consider non-equilibrium behavior of collective systems.

8

Self-Organized Criticality

*Que savons-nous si des creations de monde ne sont point
déterminées par des chutes de grains de sable?*
   *(How do we know that the creation of worlds is not determined
by sand slides?)*

Victor Hugo, Les Miserables, 1862.

Thermodynamics without an Operating Engineer

For a long time, the field of thermodynamics was strongly influenced
by thinking about steam engines. It started that way, with the considerations of the French engineer and physicist Sadi Carnot (1796–1832),
whose conclusions about the efficiency of steam engines formed the very
beginning of thermodynamics. In a steam engine, heat is supplied to a
given system, such as a gas, and the expansion of this gas is then made to
perform work. It would be ideal if all of the heat could be converted into
work, but this is in fact not possible; some of the heat has to be emitted as such, reducing the possible amount of work. The statement that
in a thermodynamic process not all of the heat can be used to do work
is in fact one way to state the second law of thermodynamics, which we
addressed in Chapter 2.

The conceptual relation between thermodynamics and steam engines
had a curious consequence, distinguishing it from other branches of
physics. Falling apples, swinging pendulums, pointing magnets, orbiting planets—they all happen on their own, they don't need an operator.
So classical mechanics and electrodynamics describe nature without any
human interference. In contrast, the steam engine needed a stoker to
fire it up, and in thermodynamics this somehow led to the picture of

someone there to raise or lower the temperature of thermal systems, or introduce an "outside" magnetic field. Critical behavior arises when an operator brings the system closer and closer to the critical point. This operator tunes the control parameter, say the temperature, and as a consequence, an order parameter, such as the magnetization or the density, shows critical behavior, and becomes singular.

In recent years, however, more and more attention has been focused on systems which can do the tuning on their own, which don't need an operator. The Danish physicist Per Bak introduced the concept of *self-criticality* for such systems, and he pointed out that in nature there are numerous instances of this behavior, in physics, biology, geology and more. The crust of the Earth slowly contracts more and more, until suddenly an earthquake releases the pressure. In Bak's favorite picture, sand falling onto a flat surface creates a sandpile becoming steeper and steeper, until suddenly avalanches start sliding down the slope. As the temperature on the surface of the Earth rises through sunshine, warm air starts moving upwards towards cooler regions. When the vertical temperature difference exceeds a certain point, turbulence sets in, creating tornados. Let us then look at self-organized criticality in more detail.

We had noted that at a critical point, one has clusters of all sizes, consisting of correlated constituents; even spins very far apart become aware of each other. Mathematically, this idea is expressed in terms of the correlation function $F(r, T)$, specifying the correlation between two constituents at separation r for a system of temperature T. This function takes the form

$$F\left(r, T\right) = \frac{1}{r^a}\, e^{-r/\lambda(T)}, \tag{8.1}$$

where $\lambda(T)$ specifies the characteristic cluster size at temperature T; the exponent a is usually close to one. The equation tells us that away from the critical region, the size of the cluster does not extend significantly beyond λ; the system has an intrinsic, though temperature-dependent scale $\lambda(T)$. As the temperature approaches the critical point, however, for $T \to T_c$, scaling sets in, λ becomes infinite, so that there is no more intrinsic scale and the correlation function reduces to

$$F(r, T_c) = \frac{1}{r^a}. \qquad (8.2)$$

This so-called *power-law* behavior is a most remarkable result. It tells us that if we compare two clusters of sizes $r = x$ and $r = y$, their relative probability $F(x, T_c) / F(y, T_c) = (y/x)^a$ is the same as that for two clusters ten times ($10x$ and $10y$) or a hundred times ($100x$ and $100y$) larger. The system becomes scale-free: there is no characteristic cluster size: all clusters obey equation (8.2). In logarithmic form, the law says that $\log F(r, T_c) = -a \log r$, so that the plot of $\log F$ vs. $\log r$ becomes a straight line.

We saw above that in equilibrium statistical systems, such power law behavior is rare. One has to have an operator tuning some parameter such as the temperature to bring the system to a critical point in order to achieve such a power law form. But in nature, this kind of behavior is found quite often. How does that happen?

A striking consequence of this arises in the study of earthquakes. Geologists classify earthquakes according to their strength on the Richter scale, from 1 to about 10, indicating the amount of energy released by the quake. (actually, the scale is open-ended, but quakes larger than 10 imply universal destruction and have so far fortunately not been observed on earth.) Previously it was often thought that the distribution of earthquakes was more or less random, with many "typical" small quakes followed once in a while by an atypical "big one." Using data collected in Missouri, USA, over a period of 10 years, Bak made a plot showing the number of earthquakes of a given magnitude during that time versus that magnitude. And surprisingly enough, the earthquakes follow the power-law of equation (6.2), as shown in Fig. 8.1. This was previously known on a purely empirical basis as the Gutenberg–Richter law—Bak showed that in fact it implies self-organized criticality. It removes the big quakes from their pedestal: in a logarithmic plot, the relative frequency of quakes of scales 1 and 2 is the same as that of scales 6 and 7. The deviation observed here for quakes of very small magnitudes is due to the fact that they are very difficult to detect. In general, scaling means that one law governs all; there is no typical quake. And the fact that the size distribution of the earthquakes has a power-spectrum implies that the

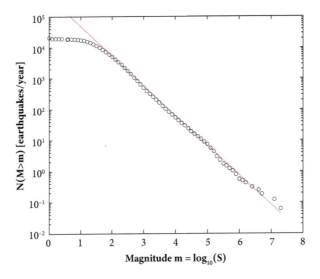

Fig. 8.1 The number of earthquakes per year, versus their magnitude on the Richter scale, in the New Madrid region of Missouri, USA.

crust of the earth has successfully adjusted itself into a critical state—no operator is needed for any tuning. It stays their on its own.

There are numerous models attempting to describe what is happening in earthquake formation. The simplest is probably the so-called slip-stick model. Imagine a block of wood placed on top of a long wooden board and subjected to a pulling spring force (see Fig. 8.2). The friction between the two wooden surfaces tends to hold the top block in place, the applied force wants to pull it in its direction. When this force becomes strong enough to overcome the friction, the top piece of wood is abruptly pulled forward. This jump reduces the spring force, and it comes to a halt again. As soon as the spring force builds up enough again, the process continues. As a result, the top piece undergoes jumps ("slips") and stops ("sticks") of different lengths in space and time.

We can now list the lengths L of the jumps and the occurrence frequency $N(L)$ for a given length, to find that

$$N(L) \sim L^{-a} \text{ or } \log N(L) \sim -a \log L. \tag{8.3}$$

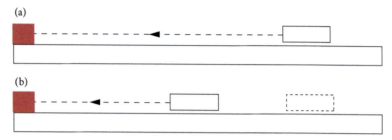

Fig. 8.2 The slip-stick model for earthquake formation.

In other words, the length distribution is found to show power law behavior, as in equation (8.2). The exponential constant a lies typically around unity.

The temperature gradient of the earth's interior places the crust under a tectonic force, giving rise to faults, on which plates strain against each other. After some time, the straining force overcome the frictional holding force, and the plates undergo an abrupt shift or slip relative to each other: there is an earthquake.

This rather simplistic picture provides the mechanism needed to understand the Gutenberg–Richter law providing a universal picture for earthquakes of all sizes, with a power-like distribution of the energy content.

In Bak's favorite example, the sandpile, the order parameter is the slope of the pile, and gravity provides the constant pulling force on sand clusters. As more and more sand is poured, the slope gets ever closer to its critical value, and in the process, an increasing number of growing avalanches is released. At the critical point, there are avalanches of all sizes, and once this point is reached, the system remains there, no matter how much more sand we add. One now records the number $N(s)$ of avalanches of size s thus produced in the course of time, and it is found that $N(s) \sim 1/s$, the power law of critical behavior. It actually turns out that this experiment is not as easy and straightforward as it sounds, since sand has the tendency to clump. The most conclusive experiments showing such behavior were eventually obtained by a group in Oslo, piling rice rather than sand.

To obtain a quantitative model, Bak and collaborators have proposed a simple grid scheme, which can be used for sandpiles as well as for earthquakes. We consider this model in more detail in the next chapter.

Further Reading

Per Bak and Kan Chen, Self-organized Criticality, Scientific American (1991) 46.
Per Bak, *How Nature works*, 1996.

9

Fractal Dimensions

*Clouds are not spheres, mountains are not cones, coastlines are
not circles, and bark is not smooth, nor does lightning travel in a
straight line.*

Benoit Mandelbrot, *The Fractal Geometry of Nature*,
Freeman, New York 1977

Dimensions

The concept of dimension arises quite naturally in our world. There
is up and down, forward and backward, left and right, so our world
is of three dimensions. Following Einstein, it is often helpful to add
time as a fourth dimension, past and future. In this world we find one-
dimensional structures, such as strings, two-dimensional ones such as
disks, and three-dimensional ones, like cubes. In physics, some theorists
find it useful to generalize and consider worlds of higher dimensions,
up to twelve and more, largely as mathematical constructs. The general
discussion of dimensions was started in 1918 by the German mathematician Felix Hausdorff; a point has dimension zero, a line dimension one,
an area two and a volume three. All these dimensions are counted by
integers, and that seems completely natural. However, recent developments have led to dimensions in between, fractal dimensions, such as
numbers between one and two, *fractals*. How that can happen?

Scaling and Fractal Behavior

Let us consider scale-free behavior in a general way. In many cases,
if we look at a jagged object at ever finer scales, it will eventually

become smooth: a staircase presents us with a rising pattern of steps, but each step on its own is quite flat. The term *fractal*, invented by the above-mentioned French mathematician Benoit Mandelbrot, is reserved for those cases where that does not happen. Both self-similar objects and more random objects can be fractal. We have already encountered the Sierpiński triangle; another quite celebrated case of a self-similar construct is the snowflake, for mathematicians the Koch curve, due to the Swedish mathematician Helge von Koch; we shall shortly return to this. A random case is given by the mentioned coastlines, as well as by Brownian motion, to be discussed in a later chapter. To deal with such general cases, we have to generalize the idea of dimension, defining it in terms of scaling behavior.

Take a straight line segment of length $L = 1$ meter and measure it first with a stick of one meter length, then with smaller ones of $1/3$ meter length. In the first case we need one stick, in the second three. In this case, the dimension D is given by the relation

$$N = s^{-D} \tag{9.1}$$

where N counts the number of sticks needed and s denotes the scaling factor, so that in the second case of above example, $N = 3$, $s = 1/3$ and hence $D = 1$. In two dimensions, squares replace the segments, giving us $N = 9$, which with $s = 1/3$ leads to $D = 2$, and in three dimensions we have $N = 27$ little cubes, so that $D = 3$. The three cases are illustrated in Fig. 9.1. Further generalizations to higher dimensions are readily obtained, Note that in each case the scaling factor, the change of the elementary length, is $s = 1/3$.

The Koch curve, the mathematical version of a snowflake, is obtained by starting with a equilateral triangle, with sides of length d. Each side is then divided into three equal parts, of lengths $d/3$, and the central segment is replaced by a correspondingly smaller equilateral triangle, of sides $d/3$ as well. This procedure is continued ad infinitum. What is length of the edge of such a construct? Initially, we had $L_1 = 3\,d$, if each side of the triangle was of length d. The next step gives twelve triangle sides, each of length $d/3$, so that the edge length now becomes $L_2 = 4\,d$.

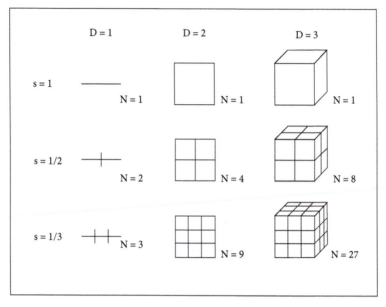

Fig. 9.1 Dimensionality vs. scaling.

This is followed by $L_3 = (16/3)\, d$. For the overall peripheral length we obtain after n steps of the relation

$$L_n = L_1 \left(\frac{4}{3}\right)^{n-1} = 3\, d \left(\frac{4}{3}\right)^{n-1} \tag{9.2}$$

so that the periphery of a finite size planar object diverges as we go to ever finer resolution, even though its area remains finite. Objects of this type were designated as fractal by Mandelbrot. The first steps of the construction pattern for the Koch curve is shown in Fig. 9.2

In the first step here, we rescale the measuring stick by one of 1/3 the original length. However, instead of getting three segments through the Koch procedure, we now get four: for $s = 1/3$, we have $N = 4$. If apply equation (8.3), we get that the dimension of the Koch curve becomes

$$D = -\frac{logN}{logs} = \frac{log4}{log3} = 1.2619 \tag{9.3}$$

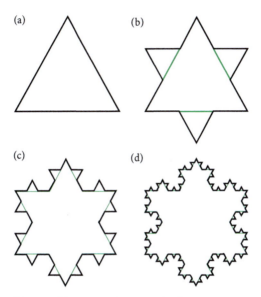

Fig. 9.2 The successive steps in the
construction of the Koch curve.

In conventional geometry, we have that the peripheral curve of a planar object is one-dimensional, its area two-dimensional. Here we find that the peripheral Koch curve is in between: it is more than one-dimensional, but less than two-dimensional.

We have here introduced a construction plan in order to arrive at fractal dimensions. In the study of self-organized criticality it was found, however, that random processes can drive the system on its own to criticality and fractal behavior. Let us see how that happens.

Consider a 3×3 grid, and place grains of sand, or little disks, or whatever, in each of the nine resulting boxes. We allow a maximum of three constituents per box; if this number is exceeded, the pile "topples," sending one particle into each of the four adjacent boxes. In other words, the size of the toppling pile is reduced by four. We continue the process until no box contains more than three objects. At the edges, the toppling elements simply fall out of the game. In Fig. 9.3, some steps of the process are illustrated. In Fig. 9.3a, we show a stable configuration, not subject to any toppling. In 9.3b, we have put four grains into the center box of a new configuration, resulting in the illustrated toppling. Finally, in Fig. 9.3c, we take still another configuration and put four grains into the center box, leading to two successive topplings.

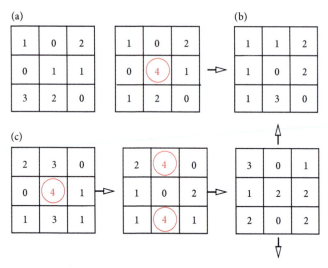

Fig. 9.3 Three different occupation configurations for a 3×3 grid.

The result of such toppling is quite counterintuitive and is due to the self-organized fractality. To illustrate this, we consider in Fig. 9.4a an empty grid of 5×5 boxes and insert into the center box 35 grains. Naively, one would imagine that the insertion leads to a pile of sand, peaked at the center and falling off towards the edges; this form is shown in Fig. 9.4b, but it is not at all what really happens. As one can readily show by playing the successive toppling steps, starting with 35 grains in the center, the final result is the one shown in 9.4c.

We can now extend this picture to a grid of arbitrary size and randomly pour a very large number N of grains onto it, into its central box. Since no box may contain more than three grains, there will in the end not be any piling up—the excess grains simply fall off the table. The final distribution this leads to is shown in Fig. 9.5: we obtain a fully fractal pattern. Here blue means empty, light blue one, yellow two and maroon three grains of sand. Again the system has driven itself to the critical point: if we continue adding more sand, the system continues to expand,

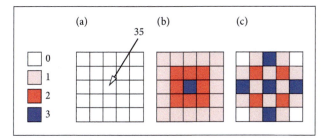

Fig. 9.4 The toppling pattern for 35 grains inserted into the central box; the form (b) corresponds to a symmetric conical sand pile, never reached. Instead, the final result is shown in (c).

retaining the same pattern, if the grid is infinite. If we do the experiment on a table of finite size, the same structure and size persists: the excess grains simply fall off the table. In other words, we can now no longer increase the size of the sandpile nor change its avalanche pattern.

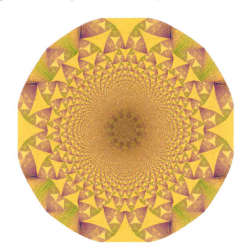

Fig. 9.5 A million grains of sand, from L. Levine, W. Pegden and C. K. Smart, arXiv:1309.3267 (2013).

Further Reading

For an exposition of fractalit, see the work of its originator,

Benoit Mandelbrot, *The Fractal Geometry of Nature*, Freeman, New York, 1977.

The pattern of the millions grains of sand is due to

L. Levine, W. Pegden and C. K. Smart, *Apollonian Structure in the Abelian Sandpile*, arXiv:1309.3267 (2013).

10

Bifurcation and Chaos

if we conceive of a being whose faculties are so sharpened that he can follow every molecule in its course, such being would be able to do what is impossible to us.

James Clerk Maxwell (1871)

The End of Determinism

The central theme of this book is that systems consisting of many constituents can and often will show a behavior which is not predictable from that observed for two constituents: that the whole is more than the sum of the parts. To define a nomenclature for this situation, let us call many-body systems simple if the whole is indeed just the sum of the parts. If it is not, but if it obeys specific new laws governing only many-body behavior, we will call it complex. If, however, the initial state of a many-body system does not allow any prediction about its future state, we will call the system chaotic. We note here that for many years, the fundamental premise of physics was that such systems don't exist.

The reason for this was the completely deterministic view of the world suggested by the time-invariant laws of classical physics. The future enters in the same way as the past, and so the great French mathematician Pierre-Simon Laplace considered a situation in which we have complete knowledge of the initial states for all constituents of the world. We can then calculate their evolution and in this way predict the future: physics gives us complete predictability. Such a view of things was later brought into doubt by the equally great French mathematician Henri Poincaré, who noted that even a minuscule error in the specification of the initial state could and would lead to dramatic changes in the evolution. So the prerequisite of the mechanistic world of Laplace, the complete specification of all initial states, was utopian, simply not achievable. Today we know that this in fact works in both directions: a tiny error in the initial state will lead to dramatic changes in both pre- and post-dictions. The climate evolution equations we might have derived do not allow us to *predict* the weather a year from now; but they also cannot *postdict* the ice age, if we run them backwards.

The Butterfly Effect

The end of determinism in this sense was probably first brought up in the nineteensixties by the American meteorologist and mathematician Edward Lorenz, professor at the Massachusetts Institute of Technology. He had formulated a set of equations specifying the evolution of weather, in term of a small number of parameters, temperature, pressure, humidity and the like. He solved these equations on a large computer, tabulating the results of a large number of iterations. At one point, he started a new set of iterations, using as input the intermediate results of a previous run. He had expected that this would give him back the subsequent results of that previous run. Instead, he found that after a few iterations, the new run would begin to deviate more and more from the prior one, and eventually, the two runs produced completely different results.

The solution to the puzzle was that the computer had recorded more than the printed number of decimal places for its future iterations, while in the new second run, Lorenz had only entered the printed numbers. In other words, the starting points of the two runs differed beyond the sixth or whatever decimal place, and that difference became crucial, became absolutely essential in arriving at the final results.

To illustrate this striking and unexpected result in general terms, this dependence on initial conditions, Lorenz noted that the flap of a butterflies wings in Brazil could in fact trigger a tornado in Texas: the subsequently famous Butterfly effect.

The Evolution of Rabbit Populations

Another essential message arising in the study of complexity is that deterministic equations do not necessarily lead to uniquely defined results.
Let us consider a well-known, albeit more biological example. As we
shall see, the evolution of the population of rabbits has contributed
significantly to our understanding of the onset of chaos—something
Australians will probably be ready to accept. If we have a population of
rabbits living on an island and consisting of N_n members in a given year
n, then an unrestrained multiplication would lead to N_{n+1} rabbits in the
next year $n+1$, with

$$N_{n+1} = kN_n \qquad (10.1)$$

where k specifies the fertility rate. For $k < 1$, the rabbits on the island
would eventually become extinguished; so from now on, we assume that
$k > 1$. Given $k = 2$, the population would double each year. If we apply
this rule to the entire island, with a starting population of 12 rabbits, this
implies 24 the next year, 48 the following, 96 in the fourth year, and so
on, always males and females in equal numbers. If we substitute the form
of equation (8.1) year by year, we obtain

$$N_n = k^n N_0 = e^{n \ln k} N_0, \qquad (10.2)$$

indicating an *exponential* explosion of the population. However, such an
increase leads to a problem: the island is presumable of finite size, and
when there are too many rabbits, there is eventually not enough food for
all. So the unbounded increase suggested by equation (9.2) becomes after
a while unrealistic. To take such a limit into account, the Belgian mathematician Pierre-François Verhulst proposed in 1838 a modified version,
the so-called *logistic map*,

$$N_{n+1} = k N_n \left(1 - [N_n/N_{max}]\right) \qquad (10.3)$$

where N_{max} specifies the maximum number of animals the island
could possibly support; in statistics, one speaks of the *carrying capacity*.

The second factor in this form implies that when that limit is reached, the rabbits have eaten all available food, so that nothing is left for survival and in the following year, there will be no more rabbits. In contrast to the unconstrained growth of equation (8.1), this form also limits the fertility rate: we must have $k \leq 4$, since otherwise the number of rabbits can exceed the carrying capacity. Let us then again assume that $N_{max} = 120$, and repeat with equation (9.3) the count we had made above, with $k = 2$. Starting again with 12 rabbits for the whole island, we now obtain 22 instead of 24 the second year, 35 instead of 42 the third and 50 instead of 96 for the fourth. The increase becomes limited in order to allow enough food for all. If we continue to iterate the process, we find that in the long run, there will be a stable average population of 60 rabbits on the island, year by year, with sufficient food for all. Moreover, it can be seen that this result does not depend on the starting number; with k and N_{max} given as above, the long term limit is always 60, no matter what the starting number is.

Obviously such a picture is oversimplified, but it does show us how a simple deterministic formulation, equation (9.3), can lead to precise predictions. To make these more general, we eliminate the dependence on the carrying capacity and divide both sides of equation (10.3) by N_{max}, to obtain

$$x_{n+1} = k x_n (1 - x_n) \tag{10.4}$$

with $x_n = N/N_{max}$ specifying the population density relative to the carrying capacity. In the above example, we had assumed a fertility rate $k = 2$ and a starting value $x_0 = 12/120 = 0.1$, leading to the sequence 0.1, 0.18, 0.30, 0.42, 0.49, 0.50.... Increasing the fertility rate, for example to $k = 2.5$, does not change the overall picture, except that in the long run (for large n) we now get $x_\infty = 0.6$, so that the limiting value now would be 72 rabbits. In Fig. 10.1 we illustrate the behavior obtained for this value of k with a starting value $x_0 = 0.5$ and compare it to that obtained for $k = 0.75$, a fertility rate less than one, with the same start. It is not surprising that in the latter case the rabbits eventually die out, since the population is reduced year after year.

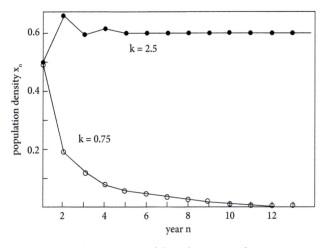

Fig. 10.1 Yearly evolution of the relative population densities x_n for fertility rates $k = 0.75$ and k = 2.5.

Bifurcation

The big and totally unexpected surprise comes when one takes k to be larger than 3. The year-by-year predictions of equation (9.4) for $k = 3.2$ is shown in Fig. 10.2. Starting as before with $x_0 = 0.5$, we obtain a strikingly

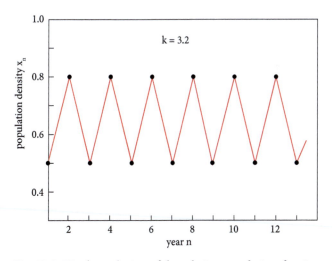

Fig. 10.2 Yearly evolution of the relative population density x_n for fertility rate $k = 3.2$.

fluctuating behavior, jumping up and down year after year and finally stabilizing at 0.5 and 0.8 in alternating years. Again the starting number does not matter—the long run result is always 0.5 and 0.8 of the maximum number, corresponding to 60 and 96 rabbits with our above carrying capacity

We thus encounter something considered impossible in a mechanistic view of the world: a fully deterministic equation and unique precise initial conditions do not lead to unique predictions. The observed behavior, with two different long run solutions, is denoted as *bifurcation* and starts when k becomes larger than 3. And it is only the beginning. For $k = 3.5$, each of the two solutions branches again, so that one now has four asymptotic values, 0.5, 0.87, 0.38 and 0.83, see Fig. 10.3. In other words: one deterministic equation, precise initial conditions, and the system fluctuates between four different predictions as soon as k increases beyond 3.45.

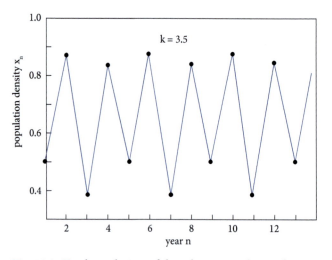

Fig. 10.3 Yearly evolution of the relative population density x_n for fertility rate $k = 3.5$.

It is in fact found that with increasing k above 3, successive bifurcations set in at an ever more rapid rate, from 2 to 4 to 8 to 16 and so on. They do so for ever smaller changes of k: the first bifurcation occurred for $k_1 = 3$, the second for $k_2 \approx 3.45$, the third for $k_3 \approx 3.54$, the fourth for $k_4 \approx 3.56$, and so on. With increasing s, the bifurcation value k_s continues to increase and gets ever closer to a limit $k_\infty \approx 3.56995....$ If we plot the behavior of the limiting value of x_n for $n \to \infty$ as function of k, we get a pitch-fork like behavior, as illustrated in Fig. 10.4, with the fork starting at $k_1 = 3$. Each arm successively forks again at $k_2 \approx 3.45$, and these forks continue the splitting, ad infinitum. In Fig. 10.4, we show only the first two bifurcations, to indicate that one can measure the linear lengths of the fork prongs, δ_1 and δ_2, as well as their linear widths σ_1, σ_2 and σ_3. Both these properties lead to limiting values.

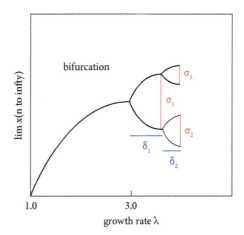

Fig. 10.4 The onset of bifurcation.

In particular, the ratio of two successive prong lengths eventually converges to a universal number:

$$s \to \infty : \delta_{s-1}/\delta_s = (k_{s-1} - k_{s-2})/(k_s - k_{s-1}) \to \delta = 4.66920 \quad (10.5)$$

This number δ is today one of the basic constants of mathematics, together with the circle measure $\pi = 3.14159...$ and the Euler number $e = 2.71828...$; it is named after the American physicist Mitchell Feigenbaum, who determined it numerically in 1975 with the help of a small pocket calculator. So the closer the fertility rate gets to the limiting value k_∞, the more bifurcations arise as solutions of the logistic map (10.2). One deterministic equation leads to arbitrarily many solutions…

The limit of the width of the forks leads to what is generally called the second Feigenbaum number, with δ being the first. The first bifurcation results in one width, σ_1 and the second to two different values, σ_2 and σ_3. The next bifurcation, not included in Fig. 10.4, leads to three distinct values, σ_3, σ_4 and σ_5.

And what happens when the fertility rate increases beyond the crucial value k_∞? Then, in the terminology used above, we get chaos, and this is why the Feigenbaum number and approach are of such importance. It implies two immediate consequences. One is that if we choose $k = 3.9$, the predicted values of N_n starting from 12 successively take on arbitrary and unordered values between zero and N_{max}:

12, 42, 107, 46, 111, 33, 92, 83, 100, 65, 116, 14, 48, 112, 28, 84, 98, …

We ask the equation how many rabbits there will be next year, and the equation answers "I don't really know, some number." The other result is that if we use a precise starting value, say exactly 12, the calculation gives the results for N_n just shown. Next, we repeat the calculations with the starting value 12.01, i.e., with a one per mil different start, and get

12, 42, 107, 46, 111, 33, 93, 82, 102, 60, 117, 11, 40, 104, 53, 115, 17, …

After about a dozen iterations, the two results become completely different. We conclude: when k lies in the chaotic regime, between $k_\infty \approx 3.56995\dots$ and $k = 4$, minute changes in the starting input number of rabbits soon lead to completely different outputs. The same effect occurs if we use the same starting input for N_n, but change the value of k by one part in a thousand: the iteration will again lead to completely different results. In other words: the result as such is not predictable, and for a miniscule change of input parameters, it will change in an unpredictable way.

This is one version of what we call chaos, and as mentioned, its appearance in this context is why the Feigenbaum scheme—at first sight a rather abstruse mathematical analysis of the logistic map—is of such fundamental importance. It signals the arrival of chaos. And as such, it shows that chaos is not some explosive collapse or end of an existing world: it is something that can smoothly arise out of the deterministic world as we know it. We can start with a fully deterministic equation, change the initial state or the input parameter by one part in a thousand, or in a million, or in whatever number: the original outcome and the new outcome will eventually become completely different. Smoothly and gradually chaos takes over.

We have here considered it in the context of rabbits on an island—simply because that is one of the most transparent situations, and the works of Lorenz and Feigenbaum were based on such problems. By now it is clear that such behavior occurs in a variety of physical situations—this best-known is perhaps the onset of turbulence, as studied by the French physicists Albert Libchaber and Jean Maurer. We will address this in a subsequent chapter. Here we will first show that rabbits were already important in mathematics several centuries ago.

Fibonacci's Rabbits: The Golden Rule

Thus we turn to another curious aspect connecting seemingly disjoint features in mathematics and science, stimulated once again by the growth laws for a rabbit population. Some five hundred years ago, the Italian mathematician Leonardi di Pisa, better known as Fibonacci (filio di Bonacci), discussed, though probably not invented, a sequence which today bears his name. He obtained it by considering in an idealized way the evolution of a rabbit population.

Starting with a juvenile pair, he assumed that after one month, the pair was mature and mated, so that after two months, an additional pair was born, one male and one female. The original pair would mate again, so that after three months, they would produce a further male/female pair, while their first offspring now reached maturity and also mated. This pattern is based on immortal rabbits and a period of one month for each newly born pair to reach maturity and mate, as well as the gestation time required after mating to produce a further pair. It leads to the evolution illustrated in Fig. 10.5 and hence to the sequence 1, 1, 2, 3, 5, 8, 13, 21, 34,... Note that each red pair has one ingoing line (its birth) and one outgoing line (its growth to maturity). In contrast, each black line has one ingoing line (its reaching maturity) and two outgoing: one for the pair it gives birth to, and one for its gestation after having mated again.

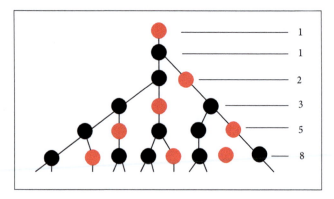

Fig. 10.5 The beginning of the Fibonacci rabbit evolution; time proceeds downwards. Red circles refer to immature (non-reproductive) pairs, black to mature reproducing pairs.

In more mathematical terms, what is now called the Fibonacci sequence is obtained by starting from tero and 1 and then successively adding the resulting integers: 0, 1, 0+1=1, 1+1=2, 1+2=3, 2+3=5, 3+5=8, 5+8=13, 8+13=21, 13+21=34, and so on, with $F_{n+2} = F_{n+1} + F_n F_n$

If we consider the ratio of two successive Fibonacci numbers,

$$2/1 = 2, \ 3/2 = 1.5, \ 5/3 = 1.66..., \ 8/5 = 1.6, \ 13/8 = 1.625, \ 21/13$$
$$= 1.615..., \ 34/21 = 1.619...$$

we find that in the limit of large numbers the ratio converges to 1.618... This value is a well-known number: it is the solution of the equation obtained if we partition a line of given length $l = a + b$ into two segments a and b such that the ratio of the sum of the segments to the larger ($a>b$) is equal to that of the ratio of the larger to the smaller: $(a+b)/a = a/b = \varphi$, see Fig. 10.6.

This is referred to the golden rule: it is considered to be the ideal ratio of two segments, and is present in many human designs, from

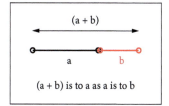

(a + b)

a   b

(a + b) is to a as a is to b

Fig. 10.6 The Golden Rule.

architecture to paintings. The equation can be solved and gives as solution $\varphi = (1 + \sqrt{5})/2 = 1.618...$ This is considered as the most irrational number of all, since it can be written as a continued fraction

$$\varphi = 1 + \cfrac{1}{1 + \cfrac{1}{1+\cfrac{1}{1+\cfrac{1}{1+...}}}} = 1.618$$

The evolution model for the rabbits used here is of course quite unrealistic: the rabbits are immortal, they reach maturity after one month and then mate, their mating always leads to one further pair, and so on—this can hardly be applied to real rabbits. To show that the Fibonacci sequence does, however, have application in real nature, we consider the genealogy of the honey bee.

The Pedigree of a Honey Bee

For bees, reproduction is a rather specialized process. The queen bee, in its mating dance, collects the sperma of numerous male bees (drones) and retains them in a special pouch of its body. When it subsequently lays eggs, it may or may not fertilize a given egg. If the egg is fertilized, it leads to a female bee, if not, to a male. Thus each male has only one ancestor, its mother, while each female bee has two, mother and father.. The resulting family tree of a male bee is illustrated in Fig. 10.7, and it is seen that the number of ancestors in each generation follow exactly the Fibonacci sequence.

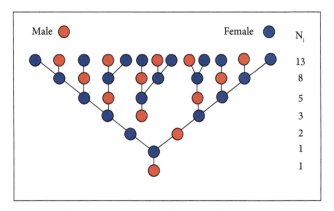

Fig. 10.7 The pedigree of a male honey bee.

In mathematics, the sequence has moreover served as a playground for numerous mathematicians, arriving at a great variety of applications. A geometric illustration in terms of plane tiling is shown in Fig. 10.8.

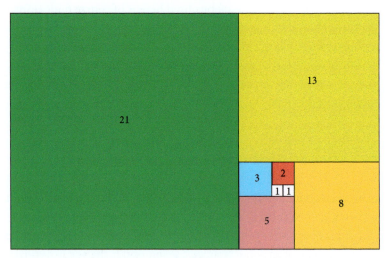

Fig. 10.8 Plane tiling according to the Fibonacci sequence.

11

Brownian Motion

*Those dust particles that we see fluctuating in sunbeams are moved
by blows that remain invisible.*

Lucretius (94–55 BC) *De Rerum Natura,*
book II, lines 114–120

Invisible Blows

In the year 1825, the Scottish botanist Robert Brown described something he had observed when he looked through a microscope at plant
pollen immersed in water. The little pollen grains did not peacefully
float there, as one would expect for quiet water, but instead they jumped
around in an irregular jittery pattern. Perhaps they were alive? He re-
peated the experiment with inorganic dust—"from rocks of all ages"—to
make sure that the motion was not somehow connected to organic life, to
an intrinsic jumping of the pollen grains themselves. It persisted. What
was the origin of this motion, today called Brownian? It took almost two
hundred years before this question was answered in the context of modern science, and the final answer was provided, so the story goes, by
none less than Albert Einstein. However, both Brown's discovery and
Einstein's explanation had predecessors, and both had noted that fact.

The phenomenon itself had actually been observed long before, even
in the days of ancient Rome. The natural philosopher (as scientists were
called in those days) Lucretius had noticed that when a sunbeam enters
a dark space, such as a barn, you would see tiny dust particles jumping
around in the air illuminated by the rays of the sun. He not only saw that,
but he also gave what we now know to be the correct explanation. The
atomic view of all matter had been developed some centuries before in

ancient Greece, and Lucretius was well aware of this atomism. In his book *De Rerum Natura* he had noted "there must be an ultimate limit to bodies, beyond perception of our senses; it is without parts and is the smallest possible thing." So to him it was quite evident what happened: the air consisted of invisible atoms, moving around rapidly and randomly, and the collisions of these invisible constituents with the larger visible dust particles caused them to jump around in the observed random way.

In the 19th century, however, the idea that all matter really consisted of atoms was heavily disputed and even among scientists not universally accepted. The great Austrian physicist Ernst Mach is supposed to have quipped, "Have you ever seen one?" And in fact it does take some degree of abstraction to picture such smooth and continuous forms of matter as air or water at rest as consisting of an immense multitude of invisible particles permanently rushing around. Opponents of atomism considered atoms at best as an artifact for calculations, not as objects that really existed. For them, atoms were somewhat like the quarks were for us in the early stages of the quark model—an idealization useful to obtain certain symmetries, but not really existing as physical objects. In any case, some direct experimental evidence for the existence and properties of atoms or molecules was called for, and it turned out that Brownian motion could indeed deliver this. When the pollen grains were put into the calm, smooth water, they jumped around, just as the dust particles did in the quiet sunlit air; and eventually, the random motion of the atoms and molecules making up the medium remained as the only explanation. The crucial step was to show that these phenomena allow us to count experimentally how many atoms (or molecules) are in a specific sample, and what the sizes of the invisible constituents are.

Following Brown's initial observation, a number of facts were recorded, concerning the motion of dust or pollen grains in water or some other liquid. The jittering particles move independently of each other, and they jump more the smaller they are—otherwise their nature and density have no effect—and they move more in a less viscous fluid (more in water than in oil). Moreover, they jitter more the higher the temperature of the medium, and the motion never stops: the particles don't get exhausted. So the scene was set…

After these observations, first a variety of non-atomic causes was considered over the years, electric effects, fluid motion, incident light, and more; they all had to be rejected. Early attempts in the budding field of statistical mechanics also failed. The collision of a single water molecule with the pollen grain gave the grain a kick, but it made it move with a speed two orders of magnitude too low. If we assume the grain to be in equilibrium with the water, so that it can oscillate and execute thermal motion as determined by the temperature of the medium, then we obtain speeds which are two orders of magnitude too high. There remained only one way out: in a given time interval, the grain, not in equilibrium with the medium, was hit on the average by more particles moving in one direction than in the others. It had to be something carried out successively by several constituents, a combined effort, complex behavior. Such an effect seemed counter-intuitive at first sight, but a second look showed it to work.

Some More Coin Flipping

If we flip a (well-balanced) coin a number of times, we expect to get in the long run an equal number of heads and tails. This result, however, does not exclude getting, for example, six heads in a row; the probability for that is in fact 1/64, so that in a hundred flips, there is indeed some chance of getting it. And after having six heads in a row, the chances of getting a tail does not increase—it remains ½, contrary to what many believe… Let us then specify a little more precisely what happens. If we throw our coin successively N times in a row, in a sequence $s(N)$ of N times, we will in general not get an equal number of heads and tails; if N is odd, this is trivially true. But if we now take a sufficiently large number R of such sequences, $s_i(N)$, with $i = 1, 2, 3, …, R$, then for each sequence with x more heads there will very likely be another with x more tails, so after averaging over all R sequences, the number of heads will equal that of tails. To avoid confusion, let us once more identify the three different numbers we used: we start with a sequence of N successive throws; we look for the difference x of heads minus tail in a specific sequence i; and finally we consider R different such sequences of N throws each.

The individual sequences will fluctuate around the average $<x> = 0$. In other words, in the set of all finite sequences of throws, there will always be a certain percentage n_k of sequences leading to $k = 2, 3, \ldots$ more heads than tails, in general not in a row, but overall in the particular sequence of throws. In the average over all such sequences (or equivalently, for one sequence with $R \to \infty$), the average number of heads becomes the same as that of tails, but any given distribution fluctuates around this average. There is no mystical operator creating sometimes k more heads (or tails) in a given sequence: the presence of such a collective effect is simply a natural consequence of random behavior, and it has its consequences. The average height of the sea level is well defined—but in a storm, there still are many waves…

In the same way, the kick on the Brownian test particle averaged over all possible water molecule configurations is zero (if there is no wind): there are on the average equally many molecules moving in any given direction as in the opposite direction. But just as the coin flipping allows a fixed percentage of sequences with k more heads than tails, so we can have in water for a given time sequence k more molecules moving in a specific direction, and such sets of k constituents persist for any value of k, albeit there are fewer the larger k is. And when such a set, for large enough k, hits the pollen grain in a given small time interval, the grain is propelled to reach a speed much greater than it could get from a single molecule. This summarizes the idea underlying the arguments presented 1906 by Einstein. The result had in fact also been found (presumably shortly before and in a different way) by the Polish theorist Marian von Smoluchowski, who had hesitated to publish them until he found that they agreed with Einstein's. In any case, although the average is zero, there are non-zero deviations. The structure is quite general and denoted today as "random walk."

To look at this phenomenon in a little more detail, we consider the simplest one-dimensional random walk. Imagine a line with points at $x = 0, \pm 1, \pm 2$, and so on, and have a test particle sit at $x_0 = 0$. At regular unit time intervals (one per second) this particle gets a kick, which moves it over by one unit, to the right or to the left, with equal probability. After the first kick, it then is either at $+1$ or at -1. The second kick puts it at $+2, 0$, or -2, the third to $+3, +1, -1, -3$, and so on. We assume that these kicks are completely uncorrelated, both from each other and from the position of the test particle. After N kicks, the displacement x_N of the test particle then is

$$x_N = k_1 + k_2 + k_3 + \dots + k_N = \Sigma_1^N k_i \qquad (11.1)$$

where $k_i = \pm 1$ specifies the value of the i-th kick. Since both $+1$ and -1 are equally likely for each kick, on the average, the result of the sum over N kicks gives a positive value and the corresponding negative value with equal probabilities. Thus we expect that

$$\langle x_N \rangle = \langle \sum_1^N k_i \rangle = 0 \qquad (11.2)$$

where < > denotes an average over a large number of samples, each consisting of a sequence of N successive kicks. An alternative way of reaching this conclusion is obtained by noting that the average over the i-th kick, averaged over many configurations, is also zero, $< k_i > = 0$ for all $i = 1$, 2, …, N, and since the kicks are independent, this means

$$< x_N > = \Sigma_1^N < k_i > = 0 \qquad (11.3)$$

These equations do not imply, however, that an individual sequence of N kicks leads to zero, nor do they tell us what percentage of the individual samples lead to a given finite (positive or negative) value of x_N, that is, what the fluctuations around the average are. We only know that positive and negative fluctuations are equally likely. Translated into the coin flipping game: the equations don't tell us how often we get six more heads than tails. We only know that six more tails will occur just as often.

To study deviations from the average, we look at something that mathematicians call *root mean square* (*rms*). We consider the average over many samples of x_N^2, given by

$$< x_N^2 > = < \left(k_1 + k_2 + \dots + k_N \right)^2 > = < \sum_1^N k_i^2 > + < \sum_{i \neq j} k_i k_j >$$

$$(11.4)$$

In the second term of the last sum, we get $+1+1$, $+1-1$, $-1+1$ and $-1-1$ with equal probabilities, making the average zero. In the first term $k_i^2 = 1$ for all i, so that there is no averaging effect and we simply have

$$< x_N^2 > = N \; \rightarrow \; r_N \equiv (< x_N^2 >)^{1/2} = \sqrt{N} \qquad (11.5)$$

for the overall rms displacement r_N after N kicks. This is the basic result of random walk theory: given a large number of sequences of N random (equal probability) steps, the root of the mean of the square of the displacement is proportional to the square root of N. Although $<x_N> = 0$, a random walk thus does not imply that we get nowhere: it just means that the region we explore in the course of all walks is given by the square root of the number of steps. We perform N steps, but the region we reach is determined by $\sqrt{N}$, while in a directed walk, we could get up to N.

After a million steps, the effective region covered is only that of a thousand linear directed steps. Since we had assumed one kick per second, N also measures the time elapsed, so that we can rewrite equation (11.4) as

$$r_t = \sqrt{t} \qquad (11.6)$$

implying that the net distance covered in a random walk increases only as the square root of the elapsed time, as a consequence of all the erratic detours.

We had here assumed steps of unit length in both space and time. More generally, equation (11.6) can be written as (dropping the subscript for simplicity)

$$r^2 = D \, t \qquad (11.7)$$

with $D = a^2 v$, where a denotes the unit step length in space and v the number of steps per second. In condensed matter physics, the quantity D becomes the so-called diffusion constant or diffusivity, defined in the study of molecular diffusion; a classical example here is the spreading of a drop of ink put into water. If we place molecules of size a (here Brownian particles) on one side of a container into a fluid of different constituents, the molecular motion of the basic fluid will in the course of time distribute the Brownian particles rather uniformly throughout the fluid:

the Brownian particles will diffuse into the fluid. Equations (11.6/11.7) constitute the first step in the analysis of Brownian motion: the rms displacement of the pollen grain is proportional to the square root of the time interval of the measurement. In a second step, the proportionality constant, the diffusivity, now has to be related to the properties of the medium: the number and the mass of the water molecules. In condensed matter physics, one finds

$$D = \frac{R}{N_A} \frac{T}{6\pi\eta a} \tag{11.8}$$

where R is the universal gas constant, η the viscosity of the medium (e.g., of water), and a the radius of the Brownian particles. The temperature and the viscosity of the liquid can be measured, and (with some difficulty) also the size a of the Brownian spheres. A measurement of the displacement r for a given time interval t thus provides us with D, and this in turn then gives us the value N_A of Avogadro's number, $N_A \approx 6 \times 10^{23}$, the number of atoms or molecules in a mole of the substance. Actually, first measurements by the French physicist Jean Perrin in 1906, shortly after the work of Einstein and Smoluchowski, gave values some 10–15% higher, but more precise values of the particle size and the viscosity led to complete agreement with other ways of determining Avogadro's number. So we still cannot see the atoms—but we know that in 16 grams of oxygen, there are N_A oxygen atoms, so that each such atom has a mass of $\frac{16}{N_A} \approx 2.7 \times 10^{-23}$ grams. Since an oxygen atom consists of 16 nucleons (8 protons and 8 neutrons), this implies a mass of 1.7×10^{-24} grams per nucleon. And we can carry out this scheme for any other medium as well, with the same result.

In the following years, the atomistic view of the world, pioneered by the ancient Greeks on philosophical grounds, became universally accepted, and in the year 1926, Jean Perrin received the Nobel Prize in Physics for experimentally confirming the atomic structure of matter.

We have here encountered one mechanism leading to an equidistribution of constituents throughout a medium: diffusion. This occurs on the level of individual collisions between the different constituents, even though successive collisions of water molecules with the Brownian particle determine the motion of the latter.

Besides such a mechanism, there are other more collective sources of motion in the medium. Imagine a balloon filled with hot air, tied to the ground: hot air implies that the molecules inside the balloon are more energetic than those on the outside. If we now puncture the balloon, the more energetic molecules from the inside will diffuse into the surrounding air. On the other hand, if we leave the balloon intact, but release it, it will rise: the hot air inside is of lower density than that on the outside—it is lighter per volume—and so the balloon will rise just as a piece of (dry) wood does in water. It is the force of buoyancy that drives it up, against the force of gravity, pulling it down: so using enough sand bags will keep it on the ground. It turns out that a fluid subject to a vertical temperature gradient shows most remarkable patterns of motion, resulting from an interplay of these different mechanisms: convection. That will be the subject of the next chapter.

12

Turbulence and Convection

Big whirls have little whirls, that feed on their velocity, and little whirls have lesser still and so on to viscosity.

Lewis Fry Richarson (inspired by Augustus de Morgan inspired by Jonathan Swift)
Weather Prediction, Cambridge University Press, 1922

Turbulence

The great American physicist Richard Feynman ranked turbulence as the most important unsolved problem in classical physics. It remains so even though it is also one of the most frequently observed phenomena whenever flow is considered, in liquids and gases of all kinds. It appears whenever the kinetic energy of a part of the fluid exceeds the damping effect of viscosity, when the moving body is tripped, so to speak. The most familiar and perhaps also the most beautiful is provided by the waves of the ocean as they reach land, so that the sand of the beach slows down the lower layers of the rushing water just above it (Fig. 12.1).

In fluid dynamics, normal behavior is laminar flow, with the fluid moving smoothly in layers above each other, with only gradual changes in velocity. In contrast, turbulent flow means irregular changes in pressure and velocity, seemingly unpredictable behavior. Nevertheless, there are situations when even turbulence occurs in regular patterns that can be described in mathematical terms, although the ultimate origin of the description still remains to be identified. Let us look at the special case of turbulent convection in more detail.

Fig. 12.1 The great wave before Kanagawa, by Hokusai (1823) National Museum Tokyo.

Convection

Convection is a consequence of the simple well-known observation that heat rises. If a fluid is heated from below, the warm bottom sections rise, while the cooler top layers sink. This occurs because the heating leads to expansion, so that the heated regions have a lower density (mass per volume) than the ones above it, and so by the principle of Archimedes, they rise upwards, lifted by the force of buoyancy. In contrast, the cooler top regions are of higher density and consequently sink. That much has been known for ages, as witnessed by Archimedes. But how does this rising and sinking occur? A volume of rising liquid—let's call it a bubble— will rub against the adjacent non-rising fluid, and this viscosity will slow down the rise. Similarly, the descent of a sinking bubble will be slowed down by viscosity. A further viscous effect will take place at the boundary of the container. It turns out that these competing effects—buoyancy and viscosity—create a structure in the fluid.

The first detailed theoretical analysis of this process was given by Lord Rayleigh in the early parts of the 20th century. He noted in particular that the temperature difference must be great enough to have the buoyancy force overcome the viscous drag, and that the heat diffusion in the medium must be slow enough to prevent the bubble from simply melting in the liquid. These different aspects can be combined to form a number, the buoyant force divided by the viscous drag and the heat diffusion. The resulting number is dimensionless and is now called the Rayleigh number R. If the other factors are kept constant, R is directly proportional to the temperature difference between bottom and top surfaces. Convection sets in when the Rayleigh number exceeds a critical value.

Let us consider the process in some more detail. If the temperature difference between top and bottom is sufficiently small, it will simply be taken care of by heat diffusion: collisions between the molecules will eventually equalize the momentum difference between top and bottom. The situation envisioned by Rayleigh assumed a large enough difference to cause macroscopic bubbles of warmer fluid to rise in the overall medium, faster than by heat diffusion, and this surprisingly enough leads to structure formation. Consider a shallow pool of fluid, say oil, in a rectangular container, with heat applied from below. In a cross section of the fluid, the rising bubbles lead to the general pattern shown in Fig. 12.2. The bubbles do not rise randomly here and there; instead, the fluid forms an array of rotating rolls, orthogonal to the longer side of the container, and of a diameter equal to the thickness of the fluid layer. The orientation of the rolls is determined such as to roll against the shorter side of the rectangle, to reduce the viscous drag. In other words, in a shallow layer of fluid, the interplay of buoyancy and drag creates a force network which leads to the array of identical, rotating, parallel rolls, whose diameter is determined by the thickness of the layer.

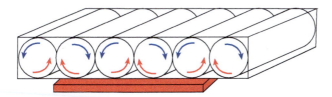

Fig. 12.2 Roll-shaped convection pattern: buoyancy vs. viscous drag.

It is known today that this scenario arises only if a further constraint is satisfied: the fluid must be contained between rigid surfaces both from above and from below. If the upper surface of the fluid is left open, this gives rise to surface tension, which in turn is temperature dependent, so that now the bubbles of hot fluid are subject to a further force in the upper surface plane, in addition to the up–down force within the medium. This creates an even more dramatic structure.

The first experimental studies of convection, carried out around 1860 by James Thomson (the brother of Lord Kelvin) in Britain and independently around 1900 by Henri Bénard in France, found that a shallow layer of open liquid when heated shows a striking tessellation pattern, similar to a mosaic of hexagons. Both scientists correctly concluded that this was the result of convection cells, but only much later did it become clear that the tessellation is a result of the interplay of surface tension, buoyancy and viscous drag.

The surface tension is lower in regions of higher temperature, and vice versa. This implies a force network covering the surface of the liquid, pulling it towards low temperature regions. The effect is illustrated for a cross section of the fluid in Fig. 12.3, to be compared to the structure with a fixed upper surface in Fig. 12.2. In the past 50 years, numerous studies have been devoted to the structure of convection in liquid layers. Combining layer thickness, buoyancy and viscous drag led to the formation of parallel rolls. Adding surface tension, the layer becomes divided into hexagonal cells. We want to emphasize here that all such structures, beyond simple heat diffusion, indicate once more that the whole is more than the sum of the parts, that there are overall collective effects determining the structure of the medium.

To convey an idea of the amazement which this phenomenon created when it was first observed, we show in Fig. 12.4 an actual surface, as it was formed in a shallow layer of silicon oil heated from below (J. A. Maroto et al., 2007). It is generally referred to as Bénard–Marangoni convection (the Italian physicist Carlo Marangoni studied surface effects of liquids in the latter 19th century), in contrast the case of a fixed upper surface, designated as Rayleigh-Bénard convection.

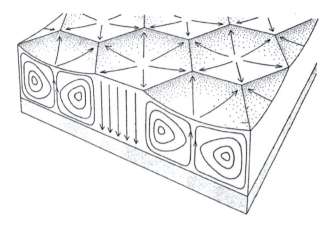

Fig. 12.3 Tessellation through convection: surface tension and buoyancy vs. drag.

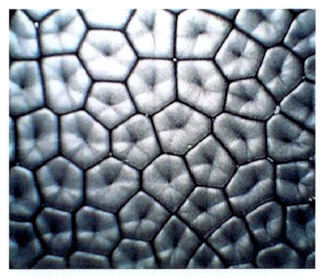

Fig. 12.4 Surface tessellation of silicon oil (from J. A. Maroto et al., Eur. J. Physics 28 (2007)).

The Road to Chaos

And that was not nearly the end of the convection story. It was continued around 1978 by the French scientists Albert Libchaber and Jean Maurer.

In a high precision experiment, they considered the behavior of liquid helium in a tiny rectangular box (width 3 mm, depth 1.5 mm, height 1.25 mm), heated from below to observe convection (see Fig. 12.5). For a box of that size, only two rolls could develop, with heat rising in the center and cooling on the sides. Thermometers in the top cover allowed a determination of the behavior of the fluid. As the bottom heating temperature was slowly increased, one first found two uniform rotating rolls, as expected from standard Rayleigh-Bénard convection: the heat was rising in the middle, between the rolls. Each thermometer plotted the temperature of the rotating role, a constant straight line as function of time.

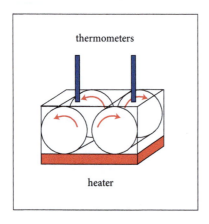

thermometers

heater

Fig. 12.5 Set-up for the Libchaber–Maurer experiment.

At a certain heating temperature value, however, each roll developed an oscillatory instability, with a wave travelling up and down its length (see Fig. 12.6).

From now on, the temperature oscillated with time in a sinusoidal pattern. The form of this pattern is one way to characterize the situation; an alternative is to register the frequency z of the oscillation, with $z = 1/\tau$, where τ is the time required for the periodic repetition, corresponding to one wave-length.

The entire experiment was carried out at temperatures close to absolute zero (around 3° Kelvin), as needed for liquid helium, and the temperature variations were correspondingly minute. As one passed a certain distance above the onset point of oscillations with a well-defined

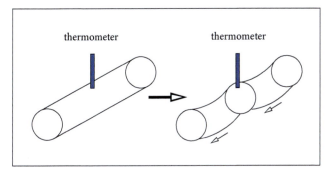

Fig. 12.6 The onset of oscillation of the convection roll.

frequency, the clear sinusoidal behavior came to an end, there appeared periodic but very irregular fluctuations. Was this the end of any theory, the beginning of chaos? To understand what was happening, we have to take recourse to the study of frequency superposition, as used for example in music.

The note A, for example, creates a sinusoidal pressure pattern in the air, with a frequency of 440 beats/second (or in standard units, of 440 hertz). The note D similarly leads to 294 hertz. If the two notes are superimposed, the resulting pattern is no longer sinusoidal, but much more complex, see Fig. 12.7 for a schematic view. Here we know that the form 12.7b is the result of superimposing two simple frequencies; but if we were just given 12.7b, how could we derive this result? In other words, how can we carry out a frequency analysis of a given time pattern?

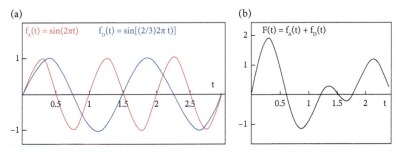

Fig. 12.7 Schematic view of the time variation of A and D notes (a) and of the sum of the two (b); for simplicity, for A we take one beat per second, for D one beat per 1.5 seconds.

The answer was given 200 years ago by the French mathematician and physicist Joseph Fourier. He proposed to transform the time distribution $f(t)$ in the form

$$p(z) = \frac{1}{2\pi} \int dt\, f(t)\, e^{-2\pi i z t}$$

to obtain the corresponding distribution $p(z)$ in terms of the frequency z; $p(z)$ is called the Fourier transform of $f(t)$. The exponential factor leads to sinusoidal oscillations, and if $f(t)$ also has pure sinusoidal oscillations, with a period a, they cancel each other, except where they are equal. Hence $p(z)$ is essentially zero, except for a peak at the oscillation frequency $1/a$ of $f(t)$. To illustrate, assume that the time distribution has the form $f(t)=\sin 2\pi a t$, with a period $1/a$. This leads to

$$p(z) = \frac{1}{2\pi} \int dt\, e^{-2\pi i z t} \sin(2\pi a t) = \frac{1}{2i} \int dt\, e^{-2\pi i t(z-a)}$$

$$= \frac{1}{2i}\, \delta(z-a),$$

where $\delta(x)$ is the Dirac delta function; it vanishes for all $x \neq 0$, and becomes infinite at $x = 0$, such that its integral is unity. We have here neglected a second term with a peak at $z + a = 0$, since we consider only positive time periods. Incidentally, $p(z)$ is pure imaginary because $\sin x$ is odd ($\sin(-x) = -\sin x$); in contrast, $\cos x$ is even ($\cos(-x) = \cos x$), and then $p(z)$ is of the same form, but real.

In Fig. 12.8 we show the Fourier transform of the pattern in Fig. 12.7b; since here $F(t)$ is the sum of two sine forms, $f_{A(t)}$ and $f_{D(t)}$, the Fourier transform shows two peaks, one at $z_A = 1$ and one at $z_D = 2/3$. And this type of result holds in general: given a complex time pattern resulting from a superposition of oscillations of a number of frequencies, its Fourier transform determines the values of these frequencies. Thus the Fourier transform was just the tool needed by Libchaber and Maurer to analyze the behavior of the oscillations of their convection rolls.

The pattern they found is shown in Fig. 12.9: the initial oscillations of the roll led to the basic frequency f_0. A further temperature increase led to a "wobble on the wobble," a second oscillation of half the initial frequency, at $f_0/2$. The time for the periodicity now doubled, from τ_0

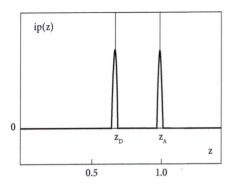

Fig. 12.8 Fourier analysis of the sum of two pure sinusoidal notes A and D, with z denoting the frequency.

to $2\tau_0$; hence the transition is referred to as period doubling. Increasing the temperature still further led to a continuation of this period doubling: there were now four peaks, at f_0, $3f_0/4$, $f_0/2$ and $f_0/4$. This constantly continued, from one peak to two to four to eight to sixteen, and so on, more and more subharmonics, with an increasing density of spectral lines. The heights of the peaks decreased at each doubling. To cite some numbers: the first oscillations (with f_0) started when the Rayleigh number (measuring the temperature difference) reached $R = 16,470$, the first bifurcation (to f_0 and $f_0/2$) for $R = 19,850$, the next (f_0, $3f_0/4$, $f_0/2$ and $f_0/4$) for $R = 20,250$), and so on. When a certain critical value $R \geq 20,750$ was reached, the periodic bifurcation structure came to an end, from now on one found a wild superposition of all frequencies: chaos.

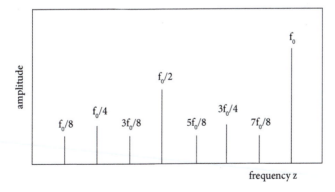

Fig. 12.9 Spectral pattern measured by Libchaber and Maurer.

When Libchaber and Maurer carried out their study, they were not aware of the results of Feigenbaum, as discussed in Chapter 9. In the summer of 1979, however, Libchaber and Feigenbaum met at a summer conference in Aspen, Colorado, and it became immediately clear that they were addressing different realizations of the same phenomenon. Feigenbaum's road to chaos, successive period doubling, was exactly what occurred in the convection experiments of Libchaber and Maurer. In Fig. 12.10, we show the spectral evolution of their experiment; the pattern evidently agrees completely with that obtained by Feigenbaum, see Fig. 10.4. And just as Feigenbaum had found that beyond a limiting fertility rate, the periodicity stopped and chaos set in, Libchaber and Maurer found that Rayleigh numbers above a critical value led to the corresponding result. In convection, temperature differences thus play the role of fertility in rabbit colonies...illustrating once more the universality aspects of complex phenomena.

The behavior obtained by Feigenbaum in an analysis of the logistic map, successive bifurcations leading to an ever more complex situation and eventually to chaos, is moreover shown to be not merely a mathematical construct. It can be observed and verified experimentally in the convection pattern of liquid helium (and by now for various other substances as well, mercury, water and more).

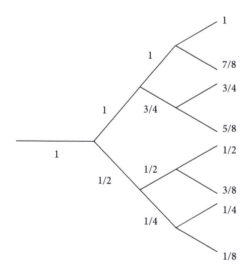

Fig. 12.10 Evolution of the spectral pattern of helium convection with increasing temperature difference; the spectral strengths are given in units of f_0.

In recognition of their achievements, Mitchell Feigenbaum and Albert Libchaber jointly received the 1986 Wolf Prize in Physics, "for opening up a totally new field of research."

Further Reading

A general introduction to convection is given in
M. G. Velarde and Ch. Normand, *Convection*, Scientific American (1960) 93.

Fig. 12.4 of surface tessellation of silicon oil is from
J. A. Maroto, V. Perez-Muñuzuri and M.S. Romero-Cano, *Introductory analysis of Bénard-Marangoni convection*, Eur. J. Physics, 28 (2007) 311.

The decisive pioneering experimental work is given in
A. Libchaber and J. Maurer, *Rayleigh-Bénard experiment in liquid helium*, Journal de Physique Lettres 40 (1979) 419.

13

Intermittency

Intermittent Flooding

In many parts of our world, natural conditions, such as the climate, develop in a more or less predictable way: it is warm in the summer, cooler in the winter, and at certain times, it may freeze or the monsoon may set in. We had already noted one exception: earthquakes, which arise in a sporadic, unpredictable way. Tornados are another instance. There are regions of the world, however, where such non-predictability is the rule. In the southwest of the USA, for example, in Arizona or New Mexico, there are numerous dry river beds, creeks or arroyos, which only carry water when—every once in a while—up in the mountains there has been a heavy rainfall. Since this happens quite rarely and irregularly, it does not make sense to build bridges wherever a road crosses such a creek. The road just passes through the creek bed from one side to the other. In case of heavy rain in the mountains, however, the quiet dusty creek bed turns into a roaring torrent, which can easily roll over a car, or even a truck, and push it downstream. To warn travelers of this danger, the crossing is framed on each side by signs such as the one at the top of this page, or others saying, "Danger, intermittent flooding." So intermittency is something which does arise in our daily lives.

In scientific terms, intermittency is usually taken to refer to the irregular occurrence of violent bursts of chaos in otherwise regular behavior—a definition which evidently agrees quite well with the flash floods in the Arizona arroyos. Is there a way to define this more precisely? Not surprisingly, intermittency shares the fate of complexity and of chaos: there does not seem to exist an unambiguous definition, and different authors use the term for quite different patterns of behavior.

It has its origin in the study of turbulence in random media, with structures (such as the "whirls" mentioned above) or peaks arising at random times in random places, with large intervals of regular flow in between (Batchelor and Townsend, 1949; Zel'dovich et al., 1987): a quiet river with unexpected rapids here and there. The structure of the universe is of this nature, with comparatively few dense clusters of matter in a vast and otherwise empty space. In meteorological terms, one can say that a thunderstorm brings intermittent strokes of lightning. Observation of a galaxy does not allow us to predict *where* the next one is located, and from watching a stroke of lightning we cannot tell *when* the next will come. From such examples we conclude that intermittent phenomena occur rarely and abruptly, off and on, and much more prominent, more spike-like than the normal pattern.

A Few Rich

To formulate this more quantitatively, we return to the example of balls in boxes or cells. Consider a container of size R subdivided into n cells of volume L ($n = R/L$) and distribute N balls into these cells. We want to see what happens if we vary n for fixed R and N, that is, when the balls are distributed into a varying number of cells. We do that once for an equidistribution of balls, the same number of balls into each cell (the normal pattern) and once with all balls into one cell (an intermittent pattern). As measure, we define the normalized moments $f_l(n)$ of order $l = 1, 2, 3...,$

$$f_l(n) = \left[\left(\frac{1}{n}\right)\sum_1^n k_m^l\right]/\left[\left(\frac{1}{n}\right)\sum_1^n k_m\right]^l$$

where k_m is number of balls in the m-th cell. The denominator is simply the l-th power of the average number N/n of balls per cell. For an equidistribution of balls, $k_m = N/n$, we find that $f_l(n) = 1$ for all n and l, as a signal of smooth normal behavior. On the other hand, if all balls are put into one cell, we have instead

$$f_l(n) = n^{l-1},$$

so that highly irregular (intermittent) behavior is reflected by a power increase of the moments as function of the number of cells. The more normally occupied cells there are, compared to the exceptional one, the more rapidly the moments increase. To illustrate that this result is not due to keeping the normal cells empty, we equidistribute half the balls, $N/2$, in first $n - 1$ cells and the other $N/2$ in the single n-th cell. A little exercise shows that then

$$f_l(n) = 2^{-l}n^{l-1}.$$

The crucial point here is that one or a few cells contain a fixed fraction of the total number of balls. This implies that the moments of a given distribution increase with the number of cells, and as a result that gives us an indication of intermittent behavior. This is in striking contrast to normal or Gaussian behavior, for which the moments are specified by the variance of the distribution, no matter how many cells are employed.

Let us see how such intermittent behavior is related to criticality in many-body systems. To have a definite case, we return to the two-dimensional Ising model, discussed in detail in Chapter 7. From the present point of view, the crucial feature is that at a given temperature the correlation between two spins i and j separated by a distance r in

general falls off exponentially, as specified by the correlation function
(see Chapter 8)

$$F_{ij}(r) = \frac{e^{-r^2/\lambda^2}}{r^a},$$

in terms of the correlation length $\lambda(T)$, which depends on the temperature T, and a critical exponent $a \simeq 1$. This implies that at a fixed
temperature, the system consists of many clusters of finite sizes $L \lesssim \lambda$.
Each of these clusters contains a finite number Q_T of aligned spins, and
in the limit of infinitely many constituents N, the ratio Q_T/N vanishes.
There is no exceptional cluster containing a fixed fraction of all N spins,
the counterpart of the cell containing a fixed fraction of the number
of balls in the above example. Only at the critical point, for $T \longrightarrow T_c$,
does the correlation length diverge, the correlation function becomes
power-like, and there appears a cluster spanning the entire system, which
contains a fixed number of all spins: the onset of criticality implies the
onset of intermittency.

The more mathematical definition of intermittency thus considers a
system of many ($i = 1, 2, ..., N$) constituents forming a large set $n \lesssim N$
of cluster states. If each of the constituents has a property q_i, we speak
of intermittency if, for at least one cluster, the sum Q of its q_i is a finite
fraction of the sum over all q_i. In other words, intermittency means that
a few rich own a fixed portion of the entire wealth of the country.

Sums vs. Products of Random Variables

We have thus far illustrated the situation by assuming specific distinct
distributions: an equidistribution of balls in a set of boxes, or one with a
fixed fraction of all balls in one box. How could such configurations arise
through self-organization in random systems? Assume N independent
random variables y_i, $i = 1, 2, 3, ..., N$, and allow each variable to take on
the values 0 or 2 with equal probability, $y_i = 0$ or 2 for all i. What are the

possible configurations of the product Y of such products,

$$Y = y_1 \times y_2 \times y_3 \times \ldots \times y_N = \prod_1^N y_i$$

It is readily seen that $Y = 0$ for all but one of the $n = 2^N$ configurations; if just one y_i has the value zero, the entire product vanishes. The exception is the case when $y_i = 2$ for all i, and then $Y = 2^N$. On the other hand, the probability for this state is $P(Y = 2^N) = 2^{-N}$. The average value of Y, averaged over all possible configurations, thus is unity,

$$\sum_n Y P(Y) = 2^N 2^{-N} = 1$$

Higher moments, on the other hand, of order $l > 1$, lead to

$$\Sigma_n Y^l P(Y) = 2^{N(l-1)} = n^{l-1}$$

and thus show intermittent behavior. This is an illustration of a quite general result: while the sum of random variables leads to a Gaussian or normal distribution, the product of such variables results in intermittent behavior.

Further Reading

Intermittency was introduced in
G. K. Batchelor and A. A. Townsend, *The Nature of Turbulent Motion at Large Wave Numbers*, Proc. R. Soc. (London) A199 (1949) 238; see also
H. K. Moffat, G. K. *Batchelor and the Homogenization of Turbulence*, Ann. Rev. Fluid Mech. 34 (2002) 19.

It was formulated as a general theory in
Ya. B. Zel'dovich et al., *Intermittency in Random Media*, Usp. Fiz. Nauk 152 (1987) 3.

For the relation between intermittency and critical behavior, see
H. Satz, *Intermittency and Critical Behavior*, Nucl. Phys. B 326 (1989) 613.

14

Words and Numbers

Fish swim, birds fly, people talk.

Norbert Hornstein
from *Noam Chomsky,* in *Routledge Encylcopedia of Philosophy,* 2017

Zipf's Law

Over the ages, humans have often wondered what distinguishes them from other animals. The answer might well be that other animals don't ask such a question. But on a more profound level, humans have the capacity to think in an abstract way: I wonder where my dead grandfather is now, or I wonder how many children my granddaughter will have? It seems that language is an essential prerequisite for such thinking. Language is the code that our built-in computer, the brain, uses to write programs.

Just as there are numerous computer codes, there are numerous languages, with assorted grammatical structures and assorted different vocabularies. Some have different genders (masculine, feminine, neuter), others not. Some express the time of an action in the form of the verb (past, present, future), others not. In some, the actor acquires a different grammatical form (nominative) than the one acted upon (accusative). No matter: all of them can convey the desired information, and all of them allow "creative" thinking about the past or the future. And no language is genetically passed on: a Chinese child growing up in a German environment will use the German form of language, and vice versa. So the computer, the human brain, is universally available, only the codes may vary and can be learned.

One obvious question appears in this context: is there some feature which all languages have in common? Simple as this may sound, there appears to be only one answer so far, known as Zipf's law, named after the American linguist George Kingsley Zipf who published, around 1936, studies on the observed regularity. It basically says this: make a ranking $r(w) = 1, 2, \ldots$, listing the most often used words w in a sufficiently large text of a given language; in English, "the" is in first place, $r(w = the) = 1$, followed by "of, and, to, a, in, is, I, that, it,..." and so on. Then register how often it was used, its frequency $n(w)$. Zipf's prediction is that

$$frequency \sim \frac{1}{rank}.$$

That is, the number of times a word is used is inversely proportional to its ranking $n(w) = A/r(w)$, where A is a universal normalization constant (independent of the second in the list, three times as often as the third, and so on. Surprisingly enough, this simple rule works in good approximation over four orders of magnitude; see Fig. 14.1 for a study using a corpus of English (Jäger and Van Rooij 2007), with some deviations

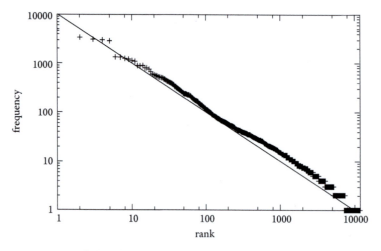

Fig. 14.1 Zipf's law for frequency and rank of words in the English language, based on the SUSANNE corpus (from G. Jäger and R. van Rooij, Synthese (2007) 159); note that both scales are logarithmic.

mainly at the beginning and the end. Moreover, it appears to hold not only in a variety of texts, from "Romeo and Juliet" to "Moby Dick," but also in all languages (Yu et al.), from English to Chinese to Esperanto; moreover, it seems to work even for those "dead" languages which are not yet deciphered. There we don't know what the words mean, but we can predict the relative frequency of their occurrence.

Zipf did not claim to have discovered this—there were a number of similar observations in quite distinct areas before him. In particular, the German physicist Felix Auerbach had noted in 1913 that if you ranked cities according to size, then the distribution of population versus rank would also follow such a law. We shall return to these, more general applications, very shortly.

In spite of much study and guessing, we don't really understand why such a law holds. In the network of our brain there may be some curious algorithm telling us to use words in such a pattern. But in the course of time, things did get even *curiouser and curiouser*, in the words of Alice in Wonderland.

The Poetry of Numbers

Zipf's law relates the occurrence frequency of elements in a set to their size. It is tempting to translate this into mathematics: take the set of all integers below 400 (as the counterpart of a Shakespeare text) and factor all integers into their prime number factors (as counterpart of words). And instead of asking what are the most frequently used words, we can now ask, what are the most frequently occurring prime numbers in the decomposition of all integers into prime numbers (a prime number is divisible only by 1 and by itself)? To illustrate: as test example we take the first integers (above 1) up to 20:

2, 3, 4 = 2 × 2, 5, 6 = 2 × 3, 7, 8 = 2 × 2 × 2, 9 = 3 × 3, 10 = 2 × 5, 11, 12 = 2 × 2 × 3, 13, 14 = 2 × 7, 15 = 35, 16 = 2 × 2 × 2 × 2.17, 18 = 2 × 3 × 3, 19, 20 = 2 × 2 × 5

note that 2, which is rank one in the set, occurs 17 times, while the next factor, 3, appears 8 times; the frequency ratio 17/8 is close to the value 16/8 = 2 predicted by Zipf's law.

Let us check the situation more generally: take the set of the first 400 integers and factorize these in terms of prime numbers. We then count the number $N(p)$ of times prime number p occurs in the set, and rank the prime numbers according to size, with $r = 1$ for 2, $r = 2$ for 3, $r = 3$ for 5, and so on. Zipf's law predicts that

$$N(p) = \frac{A}{r^x} \text{ or } log\, N(p) = log\, A - x\, log\, r,$$

with a constant A depending on the overall size of the set, and an exponent $x \sim 1$. From Fig. 14.2 we see that the power law behavior is indeed quite well satisfied, with small deviations only near $r = 1$ and in the limit of large r. The exponent is clearly larger than one; however, one finds that for increasing set size, it indeed approaches unity.

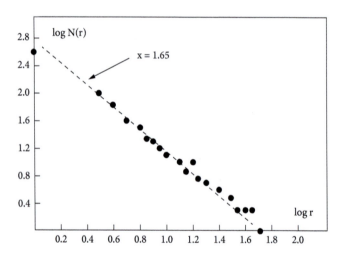

Fig. 14.2 Zipf's law for the prime factors of integers up to 400.

In Chapter 8 we had discussed the Gutenberg–Richter law, which notes that the number of times $N(s)$ an earthquake of a given strength s occurs is inversely proportional to its strength s, $N(s) \sim 1/s$. If we rank the earthquakes according to their strength, as done using the Richter scale, then the Gutenberg–Richter law becomes equivalent to Zipf's law—compare Figs. 8.1 and 14.2.

City Sizes vs. Rank

The ranking of city sizes compared to their population has just been mentioned. Already Auerbach had noted that plotting the population $N(x)$ of city x versus its rank $r(x)$ in the list of largest cities approximated the form $N(x) \sim 1/r(x)$; for a recent study, see Fig. 14.3 (with interchanged axes).

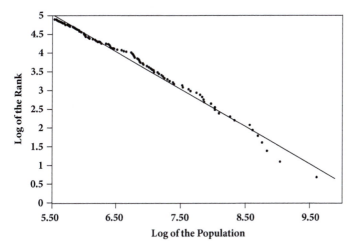

Fig. 14.3 Log size vs. log rank for the 135 largest US Metropolitan areas in 1999.

Evidently such analyses lead not only to fluctuations, but also to structural deviations. The definitions of a city size by local zoning laws are quite arbitrary, and the metropolitan region of large cities is generally much larger than the official city limits indicate. Thus metropolitan New York has some 18 million inhabitants, while the official city population is 9 million. A recent study (Bin Jiang et al.) therefore defined a "natural" size of a city by the light it emits at night in a satellite view, and then compared its rank to its size. Then the Auerbach–Zipf law was found to be much better satisfied.

To formulate the result more generally: if we are given a set of events i which we can order or rank in some way to get $r(i)$, then the number of events i, $N(i)$, follows the rule $N(i) \sim 1/r(i)$. This is perhaps the most universal law we know: it holds for earthquakes, word occurrences, city sizes, traffic jams, mathematical sequences and presumably much more. We can only ask once again: is this the result of an algorithm in the neural network of the human brain, reflecting a fundamental general pattern inherent in nature?

Further Reading

Xavier Gabaix, *Zipf's Law for Cities: An Explanation*, The Quarterly Journal of Economics, 114 (1999) 739.

Shuiyuan Yu, Chunshan Xu and Haitao Liu, *Zipf's law in 50 languages*, Computer Science arXiv 1807.01835 (2018).

Gerhard Jäger and Robert Van Rooij, *Language Structure: Psychological and Social Constraints*, Synthese (2007) 159.

Bin Jiang, Junjun Yin and Quingling Liu, *Zipf's Law for All the Natural Cities around the World.*

15

Quantum Complexity

The more important fundamental facts and laws of physical science have all been discovered, and these are now so firmly established that the possibility of their ever being supplanted in consequence of new discoveries is exceedingly remote.

Albert A. Michelson, *Light Waves and Their Uses* (1903)

The Quantum Revolution

Up to now, we have considered complexity as specific behavior shown only in the form of collective effects in systems of many constituents—as we had mentioned, a single atom cannot freeze, flow or evaporate, and it does not live in a directed time. The constituents we have looked at obey (with some slight caveats) what is now called classical physics: the science we have in mind when we think of circling planets, falling apples, cannon ball trajectories, the attraction of opposite electric charges, and more. Some 120 years ago, it was claimed by a future Nobel laureate that that is all—that physics is basically complete. The claim was already untenable when it was made...

Just as Newton's mechanics breaks down when the objects in question move too fast or become too heavy, Maxwell's electrodynamics ceases to work for very small scales, on the level of atoms and smaller. In 1905, the extension of Newtonian mechanics brought Einstein's relativity theory, and already in 1900 (i.e., before Michelson's claim) that of classical electrodynamics led to the quantum theory by Max Planck and successors. Maxwell has a moving charge radiate, but the electrons orbiting the nucleus don't. And the issue between Laplace and Poincaré, concerning to what degree the initial state of the system can be specified, was eliminated by Heisenberg: his uncertainty relations rule out that we can simultaneously determine position and momentum for even a single atom.

Initially, quantum physics indeed addressed the properties of single constituents, the energy levels of atoms, the decay of excited states, the uncertainty of complementary variables. This of course led to the question of how such quantum effects would modify the behavior of the matter made up of many individual quantum particles. Is there quantum matter, as distinct from classical matter? Are there complex quantum states, fundamentally different from complex classical states?

This question was answered in the first quarter of the 20th century on two levels, theoretical and experimental. Let us slightly deviate from chronology and start with theory. The elementary particles of present-day physics, such as nucleons, electrons and photons, can be divided into two categories, defined both by function and by properties.

On one hand, we have particles such as electrons and nucleons, which are the building blocks of atoms; on the other, we have the photons, which mediate the electromagnetic force that binds nucleons and electrons to form atoms. On an even more fundamental level, we now have quarks as the building blocks of nucleons, and gluons, binding them through the strong force. So we have some particles acting as building blocks and others as force carriers.

Quantum theory describes all particles as wave packets, detailing their distribution in space. They can thus act either as particles or as waves, depending on the specific experiment. Electrons can show wave-like diffraction patterns and photons can act as particles, as in the photoelectric effect. It turns out that the corresponding wave functions for photons are symmetric under space inversion, $\Phi(-x) = \Phi(x)$, while those for nucleons or electrons are anti-symmetric, $\Psi(-x) = -\Psi(x)$. Such symmetry behavior can be accounted for by attributing a discrete spin, a type of intrinsic rotation, to the objects in question. For the force carriers, this spin becomes integral, 0, 1 or 2; for the photons, it is unity. For the building blocks, it is half-integral: electrons and nucleons (as well as their constituents, the quarks) have spin ½.

Keeping these features in mind, we then have as one category, the so-called *bosons,* force carriers of integral spin, and as the other category, the *fermions,* half-integer spin building blocks.

The bosons are named after the Bengali physicist Satyendranath Bose, who described in 1924, at the age of 27, the behavior of a quantum gas of photons. His paper was not accepted for publication in the physics journals to which he had submitted it, so he sent it with an explanatory letter to Albert Einstein. He immediately recognized its importance, translated the paper into German and had it published in Zeitschrift für Physik, then the leading physics journal worldwide. It formed the basis of what is now known as Bose–Einstein statistics.

The fermions are named after the great Italian physicist and Nobel Laureate Enrico Fermi, who in 1926 developed the foundations for a quantum gas of such particles. Fermi was perhaps the last physicist who made seminal contributions of pioneering importance to both theoretical and experimental physics; in 1942 he led the construction of the first nuclear reactor and is therefore sometimes considered as the father of the nuclear age. His work on the properties of spin-½ particles, together with that of the British theorist Paul Dirac, is now known as Fermi–Dirac statistics, as the counterpart of Bose–Einstein statistics for particles of integer spin.

The crucial feature of fermions, which in fact formed the basis for Fermi's work, is the discovery that no two fermions with the same properties (spin, momentum) can exist at the same position in space. Fermions can never overlap, each needs its own "territory." This is encoded in the so-called exclusion principle, formulated by the Austrian theorist Wolfgang Pauli in 1925. It is this principle that results in the periodic table of elements, defining the distribution of electrons in successive concentric orbits around the nucleus, so that each electron has its own space. In contrast, nothing like this holds for bosons—they can overlap to any degree.

To obtain the spin structure of the so-called elementary particles, we have to recall that these are now considered as bound states of quarks and antiquarks, both of which have spin ½, that is, they are fermions. Mesons, such as the pion or the rho meson, are bound quark–antiquark states and thus are bosons; they have integer spins. Baryons consist of three quarks in various combinations and excitations, so that they are always fermions.

Similar considerations apply, of course, to more composite objects such as atoms, in which the spins of nuclei and electrons combine with the orbital angular momenta in a quantum-theoretical fashion. Thus hydrogen (one proton and one electron) is a boson, helium4 (two protons, two neutrons and two electrons) is a boson as well, while the isotope helium3 (two protons, one neutron and two electrons) is a fermion.

Given this information, we can now return to the issue of quantum states of matter. Classical physics considered solids, liquids, gases and plasmas as the basic forms of matter, following the proposition earth, water, air and fire made by the Greek philosopher Empedokles more than two-thousand years ago. Successsive discoveries led to subheadings, conductors, semiconductors, insulators, magnets and more. But in the course of the past century two forms of matter appeared which are purely of quantum origin, and these will form the subject in the remainder of this chapter.

The Bose–Einstein Condensate

When we lower the temperature of a gas of bosons, the corresponding wave packets become ever larger, so that for a given density they will eventually begin to overlap. The size of the atomic wave packet, its wavelength λ, is given by

$$\lambda = \frac{h}{mv}$$

where h is the Planck constant of quantum theory, m the particle mass and v its velocity. This parametrization of the particle's wave nature is due to the French theorist Louis de Broglie, and hence λ is denoted as the de Broglie wavelength. The velocity of the particle depends on the temperature of the gas, and so to reach an overlap of different particles, we need to go to extremely low temperatures. In that case, when we reach a critical temperature T_{BE}, most of the particles are in the lowest possible energy state, and then there is a percolation-like transition: the system forms one large cluster of overlapping wave functions rather than a gas of individual objects. This cluster, formed through the merger of all the constituents in the same lowest energy state, is called the *Bose–Einstein condensate*, the fifth and purely quantum state of matter, existing only at extremely low densities and temperatures, very close to absolute zero (0 kelvin = −273° Celsius).

The difficulties in achieving such low temperatures hindered the experimental search for Bose–Einstein condensates for quite some time. Only in 1995 did two groups (Eric Cornell and Carl Wieman at the University of Colorado, using rubidium atoms, and Wolfgang Ketterle at MIT, using sodium) succeed in producing the first Bose–Einstein condensates. For their achievements, they received the 2001 Nobel Prize in Physics. Since then, various other forms of this state have been produced in the laboratory.

Today we believe that some evidence of the Bose–Einstein phenomenon was in fact already known earlier. In 1938, Pyotr Kapitza in Russia and John Allen and Don Misner in England had found that below 2.17 kelvin liquid helium developed unusual liquid features: it lost all viscosity and thus became a perfect fluid, a *superfluid*. For his work, Kapitza received in 1978 the Nobel Prize in Physics.

In 1938, the German theorist Fritz London (then in Paris, in transition from Berlin to the USA, forced by the rise of Nazi power in Germany) had already suggested that the superfluidity in helium is in fact the onset of Bose–Einstein condensation, though in a fluid with rather strong interactions and not in a dilute gas, as proposed by Bose and Einstein. Helium has in fact played a crucial role in understanding the quantum states of matter. As a noble gas, its atoms interact only very weakly, and it remain in gaseous form down to −269°C (4 K), and below this point, it remains a liquid down to absolute zero. It can be solidified only under great pressure and of course at very low temperatures.

Superconductivity

The pioneering work in low temperature physics was carried out early in the 20th century by Heike Kamerlingh Onnes (Fig. 15.1) at the University of Leiden in the Netherlands. In 1908 he invented a technique to liquify helium; this required cooling it to below −269°C (4 K). He then used this liquid helium as a cryogenic tool, to cool other materials. Thanks to his rediscovered notebooks we now know that on April 8th, 1911, he measured the electrical resistance of solid mercury as a function of decreasing temperature, and to his great surprise he discovered that at −269°C (4 K) the resistance of his mercury sample fell abruptly to zero (Fig. 15.2). In the same experiment he also discovered that his cooling material, the liquified helium, became a perfect fluid at such temperatures, with zero viscosity. So this was one of the rare occasions where one experiment produced two Nobel-worthy discoveries... The Nobel committee awarded its Prize to Kamerlingh Onnes in 1913 with the wise formulation "for the investigation of the properties of matter at low temperatures."

Fig. 15.1 Heike Kamerlingh Onnes (1853–1926).

We recall that metals show an inherent conductivity: if we apply a voltage difference to a sample, an electrical current will flow. The metal consists of a lattice of atomic nuclei, which are each surrounded by an array of electrons in concentric orbits. The outermost electrons are most weakly bound and can readily move around, given a small kick by the applied voltage. In their motion, they encounter lattice impurities and other irregularities, which hinder their flow and thus give rise to electrical resistance. This resistance decreases with decreasing temperature, but it never vanishes. So what can give rise to superconductivity, to an ever-flowing current? Experiments have shown that such currents are not reduced in two years of flow—only the experimentalists became tired after that time.

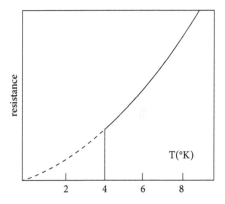

Fig. 15.2 The onset of superconductivity.

To understand the phenomenon of superconductivity a little better, we have to return to the difference between bosons and fermions. While bosons can overlap to any degree and even form a ground state Bose–Einstein condensate—one state of many fused constituents—fermions insist on having their own territory, assigned to them by the Pauli exclusion principle. This would suggest that fermions cannot form a low-temperature condensate of the kind bosons do. But nature found a way out…

Non-interacting electrons, as fermions, evidently cannot produce a condensate. But if two electrons could somehow manage to form a bound state, the two ½ spins would add up to produce a state of spin zero or one and of charge two: a boson. A similar addition takes place in the formation of, for example, hydrogen, where the proton spin and that of the electron add up to give the overall atom spin zero. But here the positive charge of the proton and the negative charge of the electron provide the necessary attraction to achieve the binding, and the final atom is electrically neutral. To have electrical flow, we need charged constituents. Moreover, two electrons *repel* each other electrically, so how could they bind to form a state of integer spin and charge two?

The puzzle was solved by Leon Cooper, who in 1956 showed that in some metals at extremely low temperatures, two electrons can indeed form a coupled bosonic state, now known as a Cooper pair. Such pairs form part of the present microscopic theory of superconductivity, proposed in 1957 jointly by the American theorists John **B**ardeen, Leon **C**ooper and John **S**chrieffer; it is now known as the BCS theory and in 1972 was honored by the Nobel Prize in Physics for the three scientists.

Cooper's approach provides us with yet another instance of complexity, of an interaction that arises as a collective effect of many constituents: alone and on their own, two electrons never feel an attractive force. Imagine a regular lattice of positive charges—the ions formed by the atoms of the metal when their outermost electrons are no longer bound. The ions are attached to their positions in a somewhat elastic fashion—they can move around a little. If now an electron passes through this lattice, it will momentarily pull towards it the closest ions in its passage (Fig. 15.3b). This creates a track of higher than average positive density in the lattice, and so subsequent electron will follow along the same path—thus behaving as if it were coupled to its predecessor, the first electron: the two form a Cooper pair (Fig. 15.3c). It's a little like a bicycle racer following closely behind a wind-breaking forerunner—the two seem coupled, a Cooper pair of bicycle racing. The effective "binding" of a Cooper pair is extremely weak, and so the phenomenon can only occur at extremely low temperatures, where the state is not immediately destroyed by lattice vibrations.

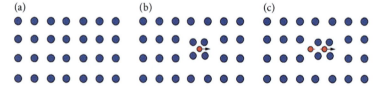

Fig. 15.3 Schematic formation of a Cooper pair.

In the right kind of metal, we thus first have normal conductivity, with a resistance decreasing as we lower the temperature. At a certain, very low critical temperature, the sudden formation of a rapidly increasing number of Cooper pairs begins, causing the overall conductivity to drop quickly to zero. We have formed a superconductor.

In a way, superconductivity is one (though a somewhat advanced) dream of mankind: one of the problems of providing energy to the general public is the loss incurred in the transmission of electrical power. Our power lines, as a result of the electrical resistance of the metallic wires, produce useless heating of the surrounding air (overground) or of the surrounding earth (underground). And where extreme values are needed, such as for the magnets of the Large Hadron Collider (LHC) at CERN in Geneva, one is already now using superconducting coils in the system of magnets. The LHC is in fact what at one time had been proposed in the USA: a superconducting supercollider, SSC. Initially, the basic problem in the use of superconductors was the fact that they functioned only at extremely low temperatures, below some 30 K. This naturally triggered a search for materials that would become superconducting at higher temperatures, allowing a wider industrial use. This search succeeded in 1986, when Georg Bednorz and Alexander Müller, working at the IBM Lab in Zürich, found that a certain class of ceramics turned superconducting around 80 K; it brought them the 1987 Nobel Prize in Physics—and the record for the shortest time span ever between discovery and Nobel Prize.

So on a quantum level, complexity has led to some rather related new states of matter; the Bose–Einstein condensate gives rise to superfluids, and with electron binding to Cooper pairs, to superconductivity.

16

Conclusion

The aim of this book was to show that an increasingly evident part of nature is complex and as such deviates from the simple laws of physics, it requires a new look and new ideas. The behavior of a large number of simple constituents leads to laws that cannot be obtained by considering two-body interactions. The whole is more than the sum of its parts.

The time flows from past to future, in spite of the intrinsic time-invariance of the fundamental laws of physics. Its direction is an emergent feature, arising from the non-equilibrium expansion of the universe. Its entropy is forever increasing, but can never reach the increasing maximum possible entropy.

The difference between actual and maximum possible entropy leads to the formation of structure as a specific form of order. Order thus becomes an unsuccessful attempt of the universe to reach total disorder.

Besides the traditional forces of physics, ultimately defined in terms to the two-body interaction between basic constituents, there are emergent forces, describing the collective drive of many particles towards a preferred state of matter. The air molecules rushing out of a punctured tire do not gain in velocity.

Gravity, long considered as the prototype of a traditional force, has never been reducible to elementary charges, and so, attempts to incorporate it into a unified description of all forces (strong nuclear, electromagnetic, weak nuclear) have failed so far. Recent suggestions claim that to be the result of the emergent nature of gravity: it is the collective effect of many particles.

Deterministic equations were generally held to produce well-defined results, provided the initial conditions were known. Today we know that the logistic map modeling the evolution of animal populations (or of turbulence in fluids) leads to bifurcation with numerous alternating solutions and eventually to chaos.

The world around us was for millennia taken as one-, two- or three-dimensional, with Einstein adding time as a fourth dimension. The study of complex geometric objects has shown that between these integer dimensions there are fractal dimensions, describing such entities as snowflakes.

In the normal behavior of matter, the communication between distant constituents quickly decreases, as described by a characteristic correlation length. At the critical transition point between two different states of matter, this length diverges and the correlation between distant members decreases as a power of their separation distance, implying an intrinsic scale invariance. Two objects at points r_1 and r_2 are thus equally correlated as two others at points 10 r_1 and 10 r_2.

A similar phenomenon is observed in seemingly unrelated issues. The frequency of words used in texts in effectively any language follows the same regularity, in linguistics known as Zipf's law: the ratio of the frequencies of the most often and second most often used words is the same as that of the 100th and the 200th most often used. In the decomposition of integers into prime numbers, the same law determines the relative occurrence frequencies, and in the study of earthquakes, the same pattern is found and known as Gutenberg–Richter law.

This shows that many features in the study of complex phenomena are of considerable universality, from prime number distributions to earthquake frequencies and word usage. The formation and structure of flocks of birds follow very similar patterns as that of magnetism in solids, and both are instances of the critical behavior and the power-law pattern at the critical point.

For millennia, we have considered four states of matter, the earth, water, air and fire of the ancient Greeks becoming solids, liquids, gases and plasmas in modern physics. The onset of quantum physics has added a fifth state of matter, the Bose–Einstein condensate. It is also defined only for many-body systems, and it leads to such striking phenomena as superconductivity (the disappearance of electric resistance) and superfluidity (the disappearance of viscosity).

In conclusion, complexity has become one of the most challenging features in today's natural science, and the development of a theoretical framework is under way—see the listed bibliography.

Bibliography

Per Bak, *How Nature Works*, Springer, New York, 199653 (1981).

Per Bak and Kan Chen, *Self-organized Criticality*, Scientific American (1991) 46.

George K. Batchelor and Albert A. Townsend, *The Nature of Turbulent Motion at Large Wave Numbers*, Proc. R. Soc. (London) A199 (1949) 238.

Predrag Cvitanović, *Universality in Chaos*, Hilger, Bristol, 1984.

Robert O. Doyle, *The Origin of Irreversibility*, The Information Philosopher 2014.

Paul Frampton, *Did Time Begin? Will Time End?*, World Scientific, Singapore 2009.

Xavier Gabaix, *Zipf's Law for Cities: An Explanation*, The Quarterly Journal of Economics, 114 (1999) 739.

James Gleick, *Chaos*, Penguin Books, New York 1987.

Brian Green, *The Elegant Universe*, W. W. Norton, New York hb1999.

John H. Holland, *Complexity: A Very Short Introduction*, Oxford University Press, 2014.

Gerhard Jäger and Robert Van Rooij, *Language Structure: Psychological and Social Constraints*, Synthese (2007) 159.

Bin Jiang, Junjun Yin and Quingling Liu, *Zipf's Law for All the Natural Cities around the World*.

Harry Kesten, *What is Percolation*, Notices American Math. Society 53 (2006) 57.

Lawrence M. Krauss, *A Universe from Nothing*, Free Press, Simon and Schuster, New York 2012.

David Layzer, *Cosmogenesis*, Oxford University Press, New York, 1990.

David Layzer, *The Arrow of Time*, Scientific American, 1975.

Albert Libchaber and Jean Maurer, *Rayleigh-Benard Experiment in Liquid Helium*, Journal de Physique Lettres 40 (1979) 419.

Benoit Mandelbrot, *The Fractal Geometry of Nature*, Freeman, New York 1977.

Melanie Mitchell, *Complexity: A Guided Tour*, Oxford 2009.

M. Mitchell Waldrop, *Complexity*, Simon & Schuster, New York 1992.

Gregoire Nicolis and Ilya Prigogine, *Exploring Complexity*, Freeman, New York 1989.

Roger Penrose, *Roads to Reality: A Complete Guide to the Laws of the Universe*, Jonathan Cape, London 2004.

Hans Reichenbach, *The Direction of Time*, U. California, Berkeley 1971 und Dover 1999.

Helmut Satz, *Intermittency and Critical Behavior*, Nucl. Phys. B 326 (1989) 613.

Helmut Satz, *The Rules of the Flock*, Oxford University Press, 2020.

Heinz-Georg Schuster and Wolfram Just, *Deterministic Chaos*, Wiley VCH, Weinheim 2005.

Dietrich Stauffer and Amnon Aharony, *Percolation Theory*, Taylor & Francis, London 1985.

Victor J. Stenger, *The Universe: the ultimate free lunch*, Eur. J. Phys. 11 (1990) 235.

Edward P. Tryon, *Is the Universe a Vacuum Fluctuation?*, Nature 248 (1973) 396.

Manuel G. Velarde and Christiane Normand, *Convection*, Scientific American (1960) 93.

Shuiyuan Yu, Chunshan Xu and Haitao Liu, *Zipf's law in 50 languages*, Computer Science arXiv 1807.01835 (2018).

Dieter Zeh, *The Physical Basis for the Direction of Time*, Springer, Berlin 1989.

Yakow B. Zel'dovich et al., *Intermittency in Random Media*, Usp. Fiz. Nauk 152 (1987) 3.

Person Index

Subject Index